Where is the Money?

Gaming Floor Innovation

By Andrew Cardno and Dr. Ralph Thomas

In Memory of Peter Mead

Peter Mead was the publisher that inspired us to write and was always there to support our latest ideas and musings. We are proud to bring this work to a book form and to continue the journey we started together. We miss you Peter.

Forward

The gaming industry is a uniquely complex industry where we essentially sell math model based experiences to customers who choose how much they would like to pay for this experience. This very creates a unique set of math problems where the normal rules of economics lack one of their primary tools, namely price. What is more the industry has come under diversified beyond gaming into a full entertainment offering. This broadening of the offering has created new analytical challenges and twists together the probability based value of customers with the actual accounting of the broader entertainment offering. It is in this environment that the authors have tackled many of the analytical challenges facing the industry.

Andrew Cardno is a software technologist serving as the Chief Technology Officer and Founder of VizExplorer, Ralph Thomas is a data scientist with a Phd in pure mathematics. They have been continuing their mad drive to publish articles in industry journals and this drive has resulted in over 100 different publications. They have worked hard to pull this body of work into a publication that links the work together and forms one of the most definitive works on gaming analytics.

"Where is the Money, Volume 1" is part one of a two part series that builds on the work in "The Math That Gaming Made". This three volume set brings together eight years of research and publication into the analytics that of the gaming industry into the book medium.

Introduction

Andrew and Ralph love math, hard problems and finding new angles to solving the analytical challenges in the gaming industry. The great challenge is not just the mathematics but the operational challenges of making these ideas practical. In this first volume we start with the core challenge of innovation and how it can be applied horizontally. We then build through the data infrastructure required to enable this innovation extend this into two primary areas: the inside space optimization challenge and how do move to a customer centric gaming model.

Each chapter is based on one published article and each was written to take the reader one step in a journey to learn about the world of gaming and how this world is moving into the realm of big data and advanced analytics. The work continues in the second volume where we shift the focus more to tackle the challenges of marketing.

Table of Contents

- CHAPTER 1: HORIZONTAL INNOVATION AND GAMING STANDARDS — 6
- CHAPTER 2: DATA AND DATABASES — 14
- CHAPTER 3: HORIZONTAL INNOVATION IN GAMING MACHINES, GAMING SOFTWARE, AND CASINO ARCHITECTURE — 21
- CHAPTER 4: GAMING DENSITY AND YIELDING THE FLOOR — 27
- CHAPTER 5: GAMING DENSITY AND YIELDING THE FLOOR — 35
- CHAPTER 6: PLAYER EXPERIENCE AND SLOT OPTIMIZATION — 43
- CHAPTER 7: FINDING THE MONEY IN JACKPOT WHARF, PART 1 — 52
- CHAPTER 8: PLAYER PREFERENCES LEARNED FROM JACKPOT WHARF, PART 2 — 60
- CHAPTER 9: BIG DATA — 69
- CHAPTER 10: BIG DATA AND LOCATIONAL INTELLIGENCE — 76
- CHAPTER 11: WAR ROOM ANALYTICS — 84
- CHAPTER 12: MAGNET GAMES AND PARADISE FISHING — 91
- CHAPTER 13: GREAT GAMES IN GAMING— WHEEL OF FORTUNE — 100
- CHAPTER 14: GREAT GAMES IN GAMING—CLUE™ — 107
- CHAPTER 15: GREAT GAMES IN GAMING— THE AMAZING BUFFALO — 117
- CHAPTER 16: OPTIMIZING PARTICIPATION GAMES — 124
- CHAPTER 17: WILL MACAU BE THE BIGGEST GAMING MARKET IN 2032? — 137
- CHAPTER 18: SERIES FINALE — 146

CHAPTER 1: HORIZONTAL INNOVATION AND GAMING STANDARDS

Authors' Note: This is part 1 of a planned 18-part series covering how money is made by connecting innovative gaming products with the customer. This article introduces the concept of horizontal products and the innovation that they bring to the whole gaming floor. When looking at horizontal innovation, we will examine two critical factors: innovation of gaming platforms and innovation of the connecting systems. We will examine a further four critical factors in subsequent articles, including innovation in analytics, innovation of gaming products, innovation in gaming hardware, and architectural innovation. This innovation is accessed in terms of how it provides or enhances the return on investment on the gaming floor, and it is compared to the vertical innovation. These areas of horizontal innovation are some of the critical features that define the player experience or shape the player's perspective of the gaming product. For this article, we welcome the contributions of Peter DeRaedt, president of the Gaming Standards Association.

Horizontal Innovation

We are defining horizontal innovation as "innovative technologies on the gaming floor that apply broadly across multiple gaming devices." These horizontal innovations are often applied to the whole gaming floor at one time. In other words, the end customer experience is in many cases a mixture of influences from many suppliers. Consider customers today. They can play a game made by one game designer, on a gaming machine made by one supplier, using a customer management experience and secondary gaming device from a second supplier, a bonusing system from a third supplier, and a ticket printer from a fourth supplier. The customer experiences the gaming device as a whole and, quite simply, has no idea that the gaming experience is provided by an amalgam of suppliers.

Game Theme
Game Platform
Customer Management System
Ticket Printer
Bonusing System

Horizontal gaming product enhancements are nothing new, and the customer management system (CMS) is a classic example of how a system that applies to every gaming product on the floor can drive enormous value.

CMS technology is a central part of how we manage customer interactions today. Harrah's was an early adopter of the information processing capabilities of the system, which eventually transformed into the Harrah's Total Rewards program. These tiered programs are now common across the industry, with considerable variety in how they are structured. This variety in how horizontal technology is implemented is in many ways the critical part of many operators' core strategies.[1,2]

Horizontal innovation opens the market to competition in many layers and reduces the barriers to entry of innovation.[3] For example, consider that a relatively small analytics company can compete with established players.[4] In this way, competition reduces the effect of "monopoly rent."[5] In other words, suppliers have to compete on price and product at every level rather than relying on infrastructure barriers to enter the market.

With the development of horizontal gaming innovation, the player experience now extends well beyond the gaming device. Therefore, for the purpose of this series, it is important to define the gaming product beyond the gaming device itself. We describe it as: "The technology and environment that defines the player gaming experience." This definition is broad enough that it includes the impact of secondary gaming devices and customer management systems.

The Role of Gaming Standards

When building horizontal technologies across the gaming floor, standards open the playing field to multiple suppliers to compete on price and product and allow these hyper-competitive companies to interoperate. This freedom of choice (what standards provide any industry) created by the Gaming Standards Association (GSA) has the promise to significantly drive innovation over the years to come.

GSA's open standards have been developed with horizontal innovation in mind. Architecturally, the Game to System (G2S) standard enables multiple hosts to co-exist on the gaming floor through the concept of owner/guest hosts. This is a key component behind the move toward horizontal technology adoption.

However, for the industry to be successful, product certification also cannot be ignored. It is the "acid test" that drives the industry toward increased interoperability, further driving more horizontal integration of multiple technologies.

Open standards allow for horizontal products, which are designed to impact every product on the gaming floor and provide additional value. Again, one of the biggest changes in the industry has been the customer management system. And, again, this system has become a standard in the industry and has changed the industry in many ways.[6] These changes are so intrinsic and fundamental that they are hard to list, but a few are free play, layered bonusing and secondary gaming devices. These changes also enable the majority of direct marketing efforts.

The Demise of the Vertical

Veteran gaming industry consultant Dr. Bart A. Lewin stated in 2008 that "Slot machines will provide a standard hardware platform providing processor(s), video, sound and player tracking capabilities. This will not only allow a multitude of game designers to build games, but I also envision the creation of high-quality game development tools and software components that have been preapproved by regulators."[7] Quite simply, he was right. The development kit exists, and horizontal innovation is happening. It is cutting deeply into vertical behavior and providing operators with choices.

In our October CEM article, "The Demise of the Gaming Manufacturer," we described how, like the computer industry, the gaming industry is likely to become horizontally partitioned. Furthermore, we have described how this change to horizontal products was accompanied by massive growth in the computer industry. Today, many horizontal companies, such as Microsoft and DellSalesforce.com, would not exist in their current forms without this shift to horizontal products.

Horizontal products such as CMS are now more important than ever in the gaming business. In many ways, the development of these horizontal products can be described as a land grab where various current suppliers to the industry, often known as manufacturers, are moving to have horizontal products. The experience of one of the authors is that the systems aspect of manufacturing is often a significant capital investment—an extremely competitive decision that in many ways defines the relationship between the "manufacturer" and the customer.

Like the software industry, this transition opens the door for dramatic change and new players to enter the market. In gaming, these new players are often enabled by standards, and their products make no attempt to compete in the vertical stack. Today's market offers more products using the standards developed by GSA.

Finding the Money

In our quest to find the money, we are committed to increasing the entire revenue of the gaming floor, taking into account the effect of cannibalization.[8] Horizontal technologies offer quite a mathematical problem when considering the measurement of incremental revenue. Quite simply, their impact is so broad the property itself provides limited access to control. In approaching the measurement of these horizontal technologies, external market measurements are often the best criterion.

Consider the example of Penny Alley, where Silverton Casino used a combination of floor reorganization and horizontal secondary device features to drive significant incremental revenue.[9] This example establishes that customer behavior can be changed by the gaming offering. Contrast the effect of Penny Alley to some new gaming products that Silverton subsequently deployed in recent months. These trial games were among the highest performing games on the floor, but calculations of their efficiency (or cannibalization) suggested no increase in incremental revenue to the property. Thus, the games were returned. Such an example shows it is possible to drive incremental revenue with horizontal innovation. In this case, the horizontal innovation now competes for resources with traditional vertical products.

Gaming Platform Innovation

The gaming platform is now standards-compliant and able to accept games from multiple suppliers. Consider properties with downloadable games: they can and do select new game themes from suppliers that have no gaming platform. Like a computer's operating system, the platform no longer controls either the content or its price.

Connecting Systems Innovation

Connecting systems are enabled by GSA standards, typically working on all gaming devices and often providing innovation in unexpected ways. Consider the coupon printer company TransAct Technologies. With its standards-compliant product line, their offerings work in machines across the gaming floor, regardless of manufacturer or gaming system.[10] This is innovation enabled by standards. As Tim Moser, former marketing manager at TransAct, once noted, "Developing on the open GSA standards and the open GSA architecture...enables operators the flexibility to leverage their existing infrastructure and provide sophisticated couponing to enhance the gaming experience in an almost limitless number of ways."

The Seminole Tribe also significantly capitalized on the adoption of GSA's open standards by using GSA's System to System (S2S) standard. It allows for connections between various Class II central servers to share player and accounting information, thus enabling player tracking and ticket-in/ticket-out across the gaming floor in a ubiquitous manner.

The Value of Horizontal Innovation

The value of horizontal innovation is not always obvious. To illustrate this point, let's compare a 3 percent lift and a 50 percent lift. Surely 50 percent is better than 3 percent, right? Perhaps not. It turns out that, when discussing horizontal versus vertical innovation, it is quite possible and indeed likely that 3 percent will be better than 50 percent.[11]

Let's assume we have discovered a horizontal innovation that provides a 3 percent lift across a large number of slot machines — let's say 25 percent of the slot floor. Compare this to a vertical innovation, such as a new penny game that provides a 50 percent lift on the bank of poor performers it replaces — a group of underachievers that only represents 2 percent of the slot floor.

Now, if we assume that 100 percent of the lift is incremental dollars, then as of right now, the horizontal innovation is not providing the same lift as the vertical innovation. The horizontal provides a 3 percent lift to 25 percent of the games, or an overall lift of 0.75 percent. (Note that this assumes all games impacted have the same average win as the rest of the floor. We will maintain this assumption throughout this example.) The vertical innovation provides a 50 percent lift to 2 percent of the floor, or 1 percent overall, which is greater than the horizontal lift.

However, when looking at the advantage of horizontal innovation (like player tracking systems) versus vertical innovation (like a brand-new game from a top manufacturer), we have to consider the concept of cannibalization.

The authors have previously discussed cannibalization in the context of optimizing a slot floor,[12] and its importance is not diminished here. Given any improvement to a slot floor, cannibalization is the percent of the improvement that came from other slot machines, and thus was not incremental. A quick example for the case of vertical innovation: Suppose we add a new game based on the new movie *Dr. T's Revenge*. This new game seems to be a hit! It does $500 in wins per day, far exceeding the $200 win per day we saw from the old poker game previously in its place. However, we already have an existing movie-themed slot machine called *Dr. T Attacks* (the predecessor to *Dr. T's Revenge*), and this game saw much of its play migrate over to the new game. In fact, *Dr. T Attacks* went from $400 WPU to $250 WPU. So, *Dr. T's Revenge* gained $300 WPU for its location, but Dr. T Attacks lost $150 WPU. In this case, Dr. T's Revenge saw 50 percent cannibalization.

With a horizontal innovation, you are improving a product that has an existing following and adding to it. Oftentimes, you are increasing things like average bet or time on device from the same customers, which means the chances for cannibalization are much lower.

With a vertical innovation like a hot, new, movie-themed penny game, you are likely to steal business from your existing movie-themed penny games. In fact, one can measure this by looking at the movement of your players. If you see significant movement from your existing game to your new game, you likely have a high cannibalization factor.

So let's return to our analysis. Let's suppose that we see 25 percent cannibalization from our horizontal innovation, meaning that 25 percent of the incremental play came from other areas of the slot floor, whereas 75 percent was incremental revenue. Let's further suppose that the vertical innovation is seen to be 50 percent cannibalization, and 50 percent incremental revenue. (In the experience of one of the authors, vertical innovations in fact tend to have a much higher cannibalization factors, often exceeding 70 percent). Our math now becomes:

Horizontal: 3% lift, 25% of the floor, 75% incremental = 0.56% lift
Vertical: 50% lift, 2% of the floor, 50% incremental = 0.50% lift

So the 3 percent horizontal lift exceeds the 50 percent vertical lift! As the slot landscape gets more competitive and cannibalization factors climb, clearly we need to look for more horizontal innovations to improve our business.

Bringing Home the Bacon
As we have shown, horizontal innovation brings home the bacon. It has been proven to do so in the past, and we think in the future it will continue to provide significant opportunities for driving value across the gaming floor. Furthermore, the traditional vertical innovation that the industry has in many ways relied on for years is now suffering from cannibalization—at least one of the author's experience, the percentage of increased revenue from game changes or the efficiency of new changes is normally less than 40 percent and often less than 10 percent.[13] In this environment, horizontal innovation may be the better option to provide operators with the opportunity to drive incremental revenue.

1. Andrew Cardno and A. K. Singh, "Who is Due Back? Part I." *Casino Enterprise Management*, August 2009, pp. 10-16
2. John Mills, "Auditing player reward programs." www.thefreelibrary.com/Auditing+player+reward+programs%3a+casino+incentive+clubs+pay+out+big...-a0162353733
3. http://en.wikipedia.org/wiki/Natural_monopoly Natural Monopoly
4. http://en.wikipedia.org/wiki/Rent-seeking
5. We are keenly aware of many small companies that compete in the analytics space, as at least one of us works for one such company.
6. John Mills, "Auditing player reward programs." www.thefreelibrary.com/Auditing+player+reward+programs%3a+casino+incentive+clubs+pay+out+big...-a0162353733
7. Cardno and Singh, "The Demise of the Slot Manufacturer," *Casino Enterprise Management*, July 2008.
8. Cardno, Singh and Thomas. "An Analyst's Guide to Slot Floor Optimization," *Casino Enterprise Management*, November 2010.
9. Cardno, Thomas and Evans. Penny Alley Series, *Casino Enterprise Management*, December 2010–March 2010.
10. The TransAct product we are referring to is sold under the brand name Epicentral ©.
11. Cardno, Thomas and Evans. Penny Alley Series, *Casino Enterprise Management*, December 2010–March 2010.
12. Cardno, Singh and Thomas. "An Analyst's Guide to Slot Floor Optimization," *Casino Enterprise Management*, November 2010.
13. Efficiency is the percentage of the total increase that is left after cannibalization is removed.

CHAPTER 2: DATA AND DATABASES

Authors' Note: This is part 2 of our "Where's the Money?" series, which covers how money is made by connecting innovative gaming products to the customer. This article examines one more critical factor in horizontal innovation: analytics. In future article parts, we will cover innovation of gaming product, innovation in the gaming hardware and architectural innovation. After we have completed these topics, we will cover how other yet-unseen forms of horizontal innovation may provide the innovation that drives growth in the gaming industry.

The emerging theme of this series is that horizontal innovation has in the past—and, we predict, will again in the future—be the core driver of enhanced profit in gaming. Furthermore, providers who are focused on horizontal products will lead this industry trend, as this is typically the nature of horizontal innovations.

Information has become one of the defining characteristics of many gaming companies. (Harrah's, for example.) One only has to look at the proliferation of analytical databases across the industry to see the impact that these information-based technologies are having. The form of the analytical database varies tremendously in a number of ways, and here we will consider three key variations: implementation technology, data structure, and data ownership.

Implementation Technology

Implementation technology can vary from MSSQL implementations to Teradata warehouse databases. These technologies, while extremely different in internal implementation, provide similar[1] Structure Query Language (SQL)[2] interfaces. However, given this huge variation in the underlying implementation, there is a dramatic similarity in that these different database platforms provide many of the same functional capabilities (but often very different performance characteristics).

Data Structure

The debates about data structure abound across the business intelligence industry. Two distinguished players with very different views here are Bill Inmon[3] and Ralph Kimball[4]. These two distinguished and highly published gentlemen have, for at least 20 years, held strong opposing views on the structure of data. It is a highly technical debate. but quite simply, one view is that all data should be held in a form with no duplication of data, and the other is that data should be held in a form that is designed for query tools. Needless to say, supporters of both views have numerous success stories supporting their views on data structure. To add even more confusion Andrew Cardno has focused on bringing data directly in the source system structure and layering in a virtualized query simplifier.

Data Ownership

There are dramatic variations between different organizational strategies when it comes to data ownership. Different categories of ownership that we have observed include:

- Third-party managed – Databases are provided and managed by a third party. Sometimes these databases can be completely outsourced to a cloud-based solution. This third party may also be responsible for delivery of analytical insights into the business.

- Industry data model – These databases apply third-party developed structures for the data to provide a framework for the initial and ongoing development of the database.

- Internal database developer managed – These databases are designed and managed by internal database developers.

In many cases, this data ownership becomes a key asset of the business. Consider the example of an operator acquiring or developing a new property. How an existing database can be applied to assist this new property could make a dramatic difference across the business, from making yield management decisions on the gaming floor to taking advantage of cross-marketing opportunities.

"Analytical innovation" can be described as extracting the most value from all the data being captured at a casino. In our opinion, in the end, many times it is driven by the level of ownership that the organization has over the data. Some data sources are more reliable than others, and the organization's ownership of the data at the source system and its dedication to improving the data's quality enables the creation of the strongest information asset. Consider the example of the hotel system in a casino business. The ability to link the data from the hotel to the player tracking system is largely dependent on the hotel management's ability to capture the player tracking number of the hotel guests. If hotel management understands the value of this data and takes ownership over its quality, then they have greater incentive, and possibly ability, to create data of higher value to the organization as a whole.

Building the Warehouse Building your Enterprise Data Asset

A typical casino has dozens of source systems, such as player tracking, gaming performance, food and beverage, hotel, and many others. The first step to maximizing the value of analytical innovation is often collecting to put all of this data into one data asset place. This data asset place carries a huge variety of names, from Reporting Server to Marketing Database to Enterprise Data Warehouse (EDW) to Realtime Virtual Dynamic Star Schema, but the defining characteristic of this data collection point is that it provides a single place to apply analytical tools and gain consistent results from data queries. For simplicity, we will refer to this single place of data storage as the Enterprise data warehouse Asset (EDAW).

Extracting the Data

Consider the role of the EDA warehouse builder. They must, in the end, extract meaningful data from the dozens or even hundreds of source systems at a modern casino.[5] The job of extracting data from source systems is a highly specialized skill that requires highly specialized tools. A non-exhaustive list of companies offering specialized tools to complete the build task includes Infomatica, Talend, Sybase, Oracle, Microsoft, and WhereScape. A vertical gaming systems company may be able to be competitive with its own source system, but without a serious realignment of its business focus, it is unlikely to compete beyond this narrow scope.

Organizing the your Enterprise Data Asset

Once the data has been extracted, it has to be put into arranged into an organized format for storage and retrieval operational decisions from the E DAW, called a database schema. There is a huge range of successful data structure strategies. In the real world of data warehousing, there is often a tremendous amount of pragmatism that is brought to bear on the structure of the database. At least one of the authors has heard phrases like "normalize until it breaks, then de-normalize until it works." This statement essentially means: follow the strict rules of data normalization6 until the performance of the database fails, then break these rules until the queries actually run.

Realtime EDA

These traditional rules of EDA building are thrown into the wind when we think about the requirements of realtime. In the realtime world seconds matter traditional data moving cost time, for example the old fashioned daily build of data is just not good enough for the realtime decisions of an operational business. For example: Consider the example of a player development tool that is showing hot players, this realtime tool needs to combine hotel, food and beverage and gaming data in realtime. The realtime tool then needs to .place this data in the hands of the host. Clearly it is not good enough to look at yesterdays data.

Analyzing the Data

Finally, the data has to make its way out of the DW EDA and into the business user's hands, and this is where a business intelligence (BI) product tools comes in.

So, data extraction, DW EDA organization, and business intelligence are three major components to the analytic process. However, we still haven't delved into the most important part: the analytics! There is great debate about where analytics should take place. In the opinion of at least one of the authors, as much analysis as possible should take place in the DWEDA. This allows for a far more robust business intelligence platform that gives the business user powerful analytics in addition to reports, rather than just providing reports and expecting the business user to do all the analytics.

Having said all this, let's go back to the idea of a vertical gaming provider attempting to provide the entire analytical process. It would have to become an expert at data extraction across dozens of source systems, provide DW EDA organization, and power DW EDA analytics for all this data.

Horizontal Analytics

Horizontal innovation is often better than vertical innovation when it comes to the gaming floor product. The same can be said of analytics. To simplify matters, let's take two source systems from the dozens that are typically available at a casino—player tracking and slot performance. Historically, the analysis of these two systems has been separate, and indeed, many casinos have two separate vertical gaming providers developing the analytics for these two data sets. We also know that marketers have made great strides in extracting value from player tracking systems, developing sophisticated direct mail programs with hundreds or thousands of segments.

For years, slot floor operators have relied on slot performance data to determine what products to place on their floors. Now imagine the power of combining these two data sets, which many casinos have already done or are in the process of doing. From the marketing side, knowing exactly what games are being played by each customer allows for improved segmentation. This information, coupled with the knowledge of new or changed slot products, marketers can reach out to their customers in far more relevant ways. The same can be said from the slot performance side. Knowing who is playing a particular game can guide product decisions. Have a handful of customers who love Game A? You may decide to keep it, even if it has below average win. Discovered a previously undetected association between Game B and Game C? Make sure to place them near each other!

Once you have the horizontal data in a DWEDA, the question becomes how best to leverage this data in a horizontal manner. One needs business intelligences tools that are "data agnostic," or, in other words, tools built for a particular player tracking or slot performance system are unlikely to be competitive against the horizontal product. The BI tool needs to be flexible enough to handle a variety of data structures.

The great debate here hinges on whether we should buy a tool off the shelf, hire consultants, or build it ourselves. Given the complexity of the analytical process, it is our opinion that off-the-shelf tools will simply never be robust enough to handle the job of horizontal analytics. Hiring consultants can and has worked—a simple Google search can reveal recent winners of DW EDA and business intelligence awards in the gaming industry. However, this is usually an expensive and time-consuming endeavor, taking roughly two to three years and millions or even tens of millions of dollars. Given the benefit of analytics, we are comfortable saying that this investment will pay off greatly for a company that goes down this road. However, it may not yet be the right option for all companies. For those with strong internal analytics, there is the option of developing it in-house.

Developing in-house tools requires some really strong human capital, from people who understand the technical complexities of an ETL EDA build to business users who can develop their own strategic concepts and communicate them effectively to the business intelligence tools developers. The good news is that the horizontal analytical process exists in layers, and these layers can be solved one at a time. For a company with the right internal capabilities, developing horizontal analytics in-house is not only a viable solution, but is often also the best solution since it gives the company the flexibility it needs to leverage analytics as a significant competitive advantage.

Bringing Home the Bacon

As we have shown, horizontal innovation and, in this case, horizontal analytics both bring home the bacon. Quite simply, the industry is being inundated with more and more systems that are required to be competitive to run the business. Horizontal analytics links all of these systems together. In many ways, horizontal analytics is the glue that holds horizontal innovation together. It is the glue that provides the operator with a single, understandable, and cohesive window into the business where insight is possible, and the outcomes of actions can be measured.

1. There are often subtle differences between SQL implementations, for example, in the date and time handling.
2. SQL is defined at http://en.wikipedia.org/wiki/SQL.
3. See www.inmoncif.com/home/ for more information about Bill Inmon.
4. See www.kimballgroup.com for more information about Ralph Kimball.
5. "The Petabyte of Gaming Data," Cardno and Singh, CEM, September 2008.
6. See http://en.wikipedia.org/wiki/Database_normalization

CHAPTER 3: HORIZONTAL INNOVATION IN GAMING MACHINES, GAMING SOFTWARE, AND CASINO ARCHITECTURE

In part 1 of this series, we briefly touched upon the power of horizontal innovation vs. vertical innovation when it comes to gaming product, and we illustrated how a smaller percentage increase in horizontal innovation is sometimes better than a larger increase in vertical innovation. In this installment, we'll give some specific examples of horizontal innovation that are available today and look into the future at what may yet be to come. First we look at horizontal innovation within a vertical.

Gaming Machines

Many slot machine manufacturers are looking at ways to improve large sections of their product line with horizontal innovation. This is not a new idea—manufacturers have long introduced new cabinet styles and then brought many of their old themes onto these new cabinets, hoping to see incremental value from all of these themes via the horizontal innovation of the new cabinet. More recent efforts following this model involve more sophisticated technology.

One slot manufacturer has released a new horizontal innovation that, among other things, allows the casino to link that slot manufacturer's games—which previously had been stand-alone and not at all connected—via a single, rapid-hit jackpot. The slot operator has indicated that simply adding this linked jackpot to its slot machines leads to double-digit percent increases in the performance of its products. At least one of the authors has verified these claims in the field, and further has noted (with no small amount of surprise) that other non-linked products from the slot manufacturer were not significantly cannibalized by the games with linked jackpots.

There are similar examples of linked jackpots that casinos have invented for their entire gaming product lines. In Las Vegas, one of the local casino giants has a jackpot that links every slot machine, and anyone playing with a rewards card is eligible for the jackpot. In addition, there is a "celebration" award so that when the jackpot hits, everyone playing at that time gets a fixed award of promotional credits. This horizontal innovation has been running for a number of years and was so successful that the company expanded it to cover games other than slot machines. While we have not seen any specific analysis to demonstrate the level of profitability of this program, it is known that this company has one of the highest percents of rated play (i.e., the percent of total play that comes from customers using their rewards card) in the industry.

Gaming Hardware

Gaming hardware is a fundamental component of the gaming offering—walk into a casino today, and you see the massive Wheel of Fortune games or the huge LCD screens with animated fish characters swimming moving on them. The great challenge with these features is that specialized hardware is oftentimes combined with the game to make unique offerings.

Gaming Platform

The gaming platform enables a vast array of capabilities, including updated themes, enhanced bonusing, and secondary gaming devices. These platforms can be enhanced with a wide range of game-specific adornments. These adornments to gaming devices provide all manner of innovations, ranging from bouncing balls to LCD reels. It is the opinion of at least one of the authors that game adornments provide for considerable visual excitement on the gaming floor.

External Systems

External non-gaming systems range from sophisticated messaging systems to music control systems. They are important, as they can provide horizontal "glue" that operates across the whole property.

Pricing Innovation

The pricing of new games is critical to the replenishment cycle of the games. Quite simply, for many operators, innovation in pricing is required to justify the new gaming machines they normally need.

Gaming Platforms

Gaming platforms are now built around gaming standards, and central to these standards is the leadership of the Gaming Standards Association (GSA). The GSA's stated mission is to "facilitate the identification, definition, development, promotion and implementation of open standards to enable innovation, education and communication for the benefit of the entire industry." It is these standards that allow the gaming device to run games from different manufacturers. This opening up of the platform creates horizontal innovation whereby game providers compete simply based on the game. While this opening up of the platform is driving competition for themes, there is also a parallel innovation whereby new games are being created with adornments or mechanical innovations.

There are innovations and challenges that arise in leveraging these new technologies for both slot manufacturers and casino operators alike. A casino floor that has deployed downloadable games technology onto every single one of its slot machines has replaced an old problem—which games should we lease/buy?—with a new, much more complicated problem—which settings should we put on every single one of those games during any given time of the week? As for the slot manufacturers, they now have to consider how they will deal with increased competition from smaller manufacturers as the ability to download games to a cabinet leads to lower and lower barriers to entry for new competitors. The level of complexity of both of these problems goes from the thousands to the billions, trillions, and beyond. It is our opinion that the real horizontal innovation isn't necessarily going to be only in the gaming product, but rather will include innovation in the analytics needed to properly optimize this new, far more complicated world.

Pricing Innovation

When looking at the insides of many gaming devices today, to quote a casino executive who preferred to remain anonymous, the components are "little more than a powerful computer with two screens." We extracted a price for a high-end gaming computer (including two screens) from the Dell website: $12,000. The price of a gaming machine, we have heard, is $25,000 or higher on the street. This leaves a $243,000 difference to allow for the construction of the cabinet and all other peripherals.

Quite simply, in this world of reduced revenues, many operators will never be able to justify a $25,000 investment, especially when games, in our experience, often have less than one year in which they outperform the house average.

Pricing innovation is seen in a number of areas, including the different participation models, prices of conversion kits, and the cost of a downloadable games library. This innovation is the start of a journey that is needed in gaming prices, and it's a journey that we hope will result in fresh gaming floors where gaming operators are able to compete for the entertainment and recreation budgets. Understanding that the competition is coming from many angles, from home entertainment to enhanced lottery offerings, is key.

External Systems

One example of external systems is a messaging system, such as SkyWire Media. According to Shawn Harris, CEO of SkyWire Media, this system enables "carefully tailored messages with specific call-to-actions to a targeted customer base that has opted-in to receive a message."

Furthermore, Harris states that "the ability to apply the SkyWire technology across the entire operations wherever a customer interacts with a POS truly drives horizontal value with our customers." This horizontal innovation is, strictly speaking, outside of the gaming system, but is designed to be integrated with the gaming experience.

Simply put, the innovation in the gaming hardware lies at the core of the gaming offering, and while innovation is often driven by periods of dramatic change, there is little doubt that the gaming product will evolve at a breakneck speed. The question is how to make it affordable to all operators and how it can be integrated with other forms of innovation to enhance the horizontal experience.

Architectural Innovation

The gaming industry has clearly been driven by architectural innovation—just look at the impact that themed properties have had on Las Vegas, where, according to Barry Thalden, "people still flock to the volcano at The Mirage, The Venetian's canals, the Bellagio's fountains and gardens, the New York New York's skyline, and the miniature Eiffel Tower at Paris Las Vegas. And they still come because gambling is fun."[1]

This architectural innovation was, for the best part of a decade, the cornerstone of innovation in the Las Vegas market. Today, however, it is in contrast with the recent building trends that have focused their innovation in other ways, as, according to Thalden, "the newer properties in Las Vegas ... have become much more sophisticated and contemporary."

In many ways, architectural innovation is the big picture. It sets the stage for the entire gaming experience—just consider the impact of the Eiffel Tower and the commitment to the distinctly French theme at Paris. This architectural innovation is, in our opinion, the cornerstone of this property's offering.

Architectural innovation has been applied in quite a different way in Indian gaming. In these properties, the innovation is often focused on the gaming floor, with large, themed designs being applied to the inside of the property. According to Thalden, "Architecture at many native casinos is also interesting and fun." The implication is that the focus of the innovation on the core business of the industry, namely the gambling product, has been successful in these environments.

Clearly, architectural innovation has been at the cornerstone of innovation in gaming in the last 10 to 20 years. While we cannot predict the future, we can say that this horizontal innovation has in the past, and is likely to again, define the gaming experience.

Horizontal Innovation Drives Revenue

The gaming industry abounds with horizontal innovations that have successfully brought home money in the past. The examples are often industry changing in their impacts; examples range from the impacts of the player tracking systems to the innovation in architecture at the Mirage in Las Vegas. As we look forward, we think the industry needs to find new horizontal innovations that can drive fun, excitement, and profit in the next 10 years. As with any innovation, it is hard to predict where it will come from, but it is likely to be a bold move by a creative group that provides this critical source of growth.

---

1. Casino Enterprise Management, January 2011, Barry Thalden, "Of Truths and Consequences: How Las Vegas Forgot How To Make Money."

CHAPTER 4: GAMING DENSITY AND YIELDING THE FLOOR

Authors' Note: This series discusses how money is made by connecting innovative gaming products to the customers. This installment looks at the density of gaming machines and how to determine if changes in this density result in increased or decreased revenue. Future article parts will explore the problem of multiple (and disparate) data streams and discuss options for maximizing the value of this kind of data. We also plan to share some real-world examples of this optimization. Think of it as optimizing incremental lift per analytical hour!

In today's gaming world, there is vastly increased game flexibility and supply of gaming products in many, or possibly most, markets. These factors are changing the rules of yield on the gaming floor, necessitating a new framework for optimizing the yield of these games in this new environment.

The term "yield management" is "used in many service industries to describe techniques to allocate limited resources, such as airplane seats or hotel rooms, among a variety of customers, such as business or leisure travelers. By adjusting this allocation, a firm can optimize the total revenue or "yield" on the investment in capacity. Since these techniques are used by firms with extremely perishable goods, or by firms with services that cannot be stored at all, these concepts and tools are often called perishable asset revenue management."[1]

In the gaming machine space, we define yield management as techniques to allocate gaming resources among a variety of patrons. For example, offering different times on a slot machine to different tiers of patrons. One measurement of this yield is revenue per square foot. Let's explore revenue per square foot, then set a framework for optimizing the yield of the gaming floor and maximizing this revenue.

The Model
In the modern casino, there are dozens of revenue streams. Casinos make money from hotel rooms, entertainment, retail shops and, of course, gambling. The question of how to optimize these multiple revenue streams traditionally in many instances has come down to the revenue-per-square-foot model.

The model is quite simple. Just figure out the daily revenue per square foot (RSF) of each revenue stream and make sure you allocate the space to ensure the highest yield. At this point, you might declare, "If that were the case, then every casino would simply allocate all their space to the revenue stream with the highest RSF!" And you would be correct ... if RSF were a constant. But it's not.

Consider the slot machine. For the sake of mathematical simplicity, let's assume that each slot machine takes up 1 square foot and has the ability to collect $100 per hour of play. Let's also assume that there is no competition for the casino we are examining, and that there are plenty of local gamblers itching to play.

If the casino were to open with exactly one slot machine, it's safe to assume that the machine would be played 100 percent of the time. The RSF for that machine would therefore be $2,400 ($100 per hour x 24 hours / 1 square foot). (See Table 1.) Two machines would also be played 100 percent of the time, and thus the RSF for two machines is still $2,400 ($2,400 x 2 machines / 2 square feet). In fact, this pattern will continue until we finally place enough slot machines that demand subsides to provide something less than 100 percent utilization. At this point, the RSF would then slowly decrease as we continue adding more games. For this example, let's assume this occurs at 100 machines.

Now suppose we've continued to add games until we've reached the point of saturation, that is, the point at which adding games ceases to add any total revenue from all slot machines combined. For the sake of this example, let's assume saturation occurs at 10,000 machines and that our RSF is $120 per day, meaning we are making

$1.2 million per day from all machines combined ($100 per square foot x 1 square foot per machine x 10,000 machines). If we were to add another 10,000 machines, our overall win per day would not change due to already being at the point of saturation, but now our RSF plummets to $60 ($1,200,000 per day / 20,000 machines).

Table 1 – Revenue Per Square Foot Calculations

No. of Slot Machines	1	2	100	10,000	20,000
Revenue Per Hour Per Machine	$100	$100	$100	$5	$2.50
Total Revenue Per Hour	$100	$200	$10,000	$50,000	$50,000
Total Revenue Per Day	$2,400	$4,800	$240,000	$1,200,000	$1,200,000
No. of Square Feet2	1	2	100	10,000	20,000
Observation/ Comment	Incremental revenue generated	Incremental revenue generated	Decreasing marginal revenue from this point on	Saturation point - no additional revenue	Saturation point - no additional revenue
RSF - Per Hour	$100	$100	$100	$5	$2.50
RSF - Per Day	$2,400	$2,400	$2,400	$120	$60

So, instead of RSF being a constant for this particular revenue stream, it is in fact a complicated, piece-wise step function—it starts off being constant, then quickly begins a slow descent when demand is met until we reach saturation, at which point it finally plummets toward zero. While rarely calculated precisely, the fact that casinos are not simply 100 percent devoted to the top revenue stream shows that operators have (at least intuitively) known for years that RSF is a highly variable metric. To unlock understanding of this problem, we need to look deeper into yield management and defining gaming patrons.

Patrons vs. Customers

Patrons are dramatically different from customers. A customer is defined as "a person who buys,"3 with the assumption that there is a seller and a price. Patrons are different for at least three reasons:

1. Patrons choose their own price for their gambling experience. The gambler chooses how much to wager and so controls the price of their experience; they choose to spend pennies on a game or tens of thousands of dollars on that same game.
2. While the house advantage underpins the whole industry, it is always possible for patrons to win, resulting in a negative purchase price. In a retail situation, the customer always pays.
3. Some patrons are actually skilled enough that they gamble with the odds in their favor.

When looking at yield management in the hotel industry, we look at business customers vs. leisure customers. When looking at gaming patrons, we need to consider other factors, as there are very few business customers and typically we try to stop them from playing at all. When looking at gaming, we need to consider the constraints that drive a patron's gaming experience. These constraints include:

1. Time – Some patrons are limited in the total amount of time that they can spend gaming.
2. Wallet – Some patrons are limited in the total amount they can lose during their gaming experience.
3. Time of day – Some patrons can only play at specific times of the day.
4. Day of week – Some patrons can only visit on specific days of the week.

In yield management, there are a number of metrics that can and possibly should alter the way that we execute our plan. However, how these metrics are normalized can provide quite different views of the patron. Normalization is defined as the denominator in the calculation of the average or rate. Following are examples of different time-based normalizations:

1. Per Hour of Play (PH)
2. Per Day (PD)
3. Per Month (PM)
4. Per Trip (PT)

To illustrate the use of normalization, consider two patrons who have the same theoretical win per trip but dramatically different theoretical win per hour of play. While their trips have basically the same value to the business, their customer profiles are vastly different, as is their suitability for yield management.

Combining the patron constraints (as discussed above) with a ranking of the time normalized metrics generates a powerful breakdown of customers suitable for yield management. Table 2 shows a breakout of players by their yield management types. For illustrative purposes, it may help to imagine that all of these players have the same theoretical win per trip and the same theoretical win per month. They do, however, present very different yield management challenges. Consider the aggressive time constrained players. It is likely important that members of this segment find the gaming experience they are looking for in small amounts of time.

Table 2

Players who have the same theo win per trip and the same theo win per month, but have:		
	Low Theo Win Per Hour	High Theo Win Per Hour
Time-constrained patrons	These patrons spend at a low rate and have limits on the amount of time they can spend at the property.	These aggressive patrons play hard and only stay for short periods.
Time-unconstrained patrons	These players play for extended periods and at a low rate.	These patrons stay for extended periods and spend at a high rate.

Yield Management Actions

- Closing whole sections of the gaming floor
- Changing secondary gaming effects
- Time-of-day marketing events
- Promotional offers, such as meals that are available only at a specific time of day
- Changing the density of gaming machines on the floor
- Changing the mix of gaming of machines
- Changing the available patron-selected options, such as theme or denomination

Beyond Gaming Machines

We've seen that the gaming yield management model is in practice much more complicated than it seems at first. But don't worry, it's actually much worse!

The various revenue streams do not exist in isolation. Instead, certain streams serve to support others. In heavily competitive environments, companies have even resorted to loss leaders— for example, a buffet that loses money but is a great marketing tool for driving incremental gaming revenues. At this point, the gaming yield optimization model falls apart completely, but do not be discouraged, analysis is still possible.

When confronted with a low- or negative-yielding revenue streams, one must answer the following questions:

1. Does this revenue stream prop up other revenue streams, like the buffet example above?
2. If so, how much incremental revenue does it provide?
3. Does the ROI on the low-performing revenue stream provide justification for the low- or negative-performing square footage?

Question 1 is really for operators, and usually the answer is a simple "yes" or "no".

Question 2 requires some heavy analytics. However, simple moving averages can often provide a gauge as to the effectiveness of one revenue stream on another.

Question 3 is quite interesting. To properly calculate the ROI when we allow square footage to be used for underperforming revenues, we cannot assume that if it's not a loss-leader, there is no cost (or if it is a loss-leader, that the cost is simply whatever amount we lose per day). Instead, we need to measure the opportunity cost. If there is a better use for that space, then that has to be factored into our ROI calculation.

An example might make this clearer. Suppose that we have a 10,000-square-foot buffet that loses $1 per square foot per day. However, we've been able to measure its impact to the gaming floor and have discovered that it adds $15,000 per day in incremental gaming revenue. So the 10,000-square-foot buffet loses $10,000 per day but adds $15,000 per day in gaming revenue for a net gain of 50 cents per square foot per day. Sounds good, right? But what if we had another use for that space that generated $1 per square foot in revenue? Or if we could spend $10,000 more per day on marketing and thereby generate $20,000 per day in incremental gaming revenue? In either case, we may want to remove the buffet after all.

Final Thoughts

We have explored the optimization of various products on a casino's footprint. We have looked at both non-gaming and gaming revenues, and have explored different analytical techniques for attacking this optimization problem. Never before has the gaming floor been so dynamic and never has the customer had so many choices from both the gaming industry and its non-traditional competitors.4 Moving to a yield managed gaming business gives a fascinating twist on the operation of the gaming experience, and like the results of yield management in other industries, may result in what seems to be counterintuitive actions. However, these counterintuitive actions could become the next competitive advantage.

1 http://archive.ite.journal.informs.org/Vol3No1/NetessineShumsky/
2 Based on the assumption that each slot machine takes up 1 square foot.
3 http://dictionary.reference.com/browse/customer
4 P. Laube, M. de Berg, M. van Kreveld (2008). Headway in Spatial Data Handling (Eds. Anne Ruas, Christopher Gold), Lecture Notes in Geoinformation and Cartography Series, pp. 575–593.

CHAPTER 5: GAMING DENSITY AND YIELDING THE FLOOR

Authors' Note: In our last article part, we focused on the best use of gaming space from the casino's perspective. This article part continues to explore how gaming analytics and the right initiatives can drive incremental revenue by looking at a real-world example of how changes in the density of gaming machines impact game performance. In future article parts, we will explore how we can use mini casinos combined with new metrics, such as corrected theoretical win, to explore the displacement of revenue from these insight-driven initiatives.

Historically, casinos have held to the theory that more is more. That is, the more games you can place on the floor, the more money you can make as a casino operator. In capacity-constrained environments, this is largely true, since the significant demand by a large number of customers justifies the additional gaming devices, even if the added devices cause the customers to have a lesser experience. The math on this is simple. Suppose we have 1,000 units and 10,000 customers spending $50 per day. Our win per unit is $500, and our total win per day is $500,000. Now let's say we add 200 units and, in doing so, are able to add 2,000 more customers. However, some of the customers reduce their play by virtue of feeling "cramped," in turn reducing our customer spend per day to $45. Now we have 1,200 units and 12,000 customers spending $45 per day. In this case, our win per unit declines to $450, but we don't care because our win per day has increased 20 percent to $540,000.

Many casinos, however, have realized that this model breaks down as competition increases and demand for their particular casino declines. We want to explore the best use of gaming space in this reduced-demand environment, specifically focusing on the customer. Quite simply, competition is driving down margins, and you don't have to look further than the aggressive marketing programs of late to understand that margins are being pressed. These margins are leading us to look at a number of critical ways to improve profit, including yield management, the correction of theoretical win allocation and its effect on bonus (free play) offerings.

Corrected Theoretical Hold

The theoretical hold percentage has always been a marginal calculation when looking at the player experience. As we've said before, "players' behavior results in quite different play effects during the game, while maintaining the same expected value (EV) EV."[1]

When we add the impacts of different pay tables on the game machine—especially, but not limited to, multi-game machines— it is a mathematical truth that the theoretical win calculation for both the players and the machine are systematically inaccurate. This inaccuracy leads to the controversial argument that the free play allocation made by many operators is systematically incorrect.

Now, before we examine a real-world example from Silverton Casino, where a reduction in density was part of a strategy to increase the yield on the gaming floor, let's explore some of the theoretical basis behind yield management.

The Infinite Gaming Floor

First, let's suppose that we have as much gaming space as we need—essentially, an infinitely large gaming floor—and we want to know how much space to place between our games. For simplicity, we will assume that we are dealing only with slot machines. In this hypothetical example, we do not have to worry about supply (we're assuming we have as much as we want/need) or demand (since demand is relative to supply and we can always adjust the supply per our liberal hypothesis). Our goal is to isolate how space between games influences customer spend, from the customer perspective.

At one extreme, we space the games so far apart that we effectively provide one game for each customer. In this extreme example, from the customer perspective, once they make their way to a game, the next game is so far away it cannot be reached in a reasonable amount of time. Imagine now that you are the customer, and all you are provided is a single slot machine! It is highly likely that you will not play all of your gaming wallet (if any at all) on that machine, and that you are not going to return to this fantastical casino any time soon.

At the other extreme, we place the games on every available square foot of space. There are no aisles, as games are placed in long rows with chairs touching each other, and the customer doesn't have enough room to move between games. As with our first extreme example, as a customer you are likely to be very unhappy with your experience and unlikely to return.

So, what is the optimal spacing from the customer perspective? That is, what density of the gaming floor will cause the customer to have the best experience and maximize the likelihood that the customer will play all of their gaming wallet at that casino? Considering our two extreme examples, and recognizing that the answer lies somewhere in between, we end up with a sort of "Laffer Curve" for gaming density (see Figure 1).

Figure 1: Laffer Curve for Gaming Density

While Arthur Laffer argued that the optimal taxation rate lied somewhere between 0 percent and 100 percent2, we argue that the optimal gaming density for a customer lies somewhere between zero (where the games are stacked as closely together as possible) and infinity (where the games are so far apart as to be unreachable by the customer). Our goal is to find the top of the curve.

Now, in practice, our hypothetical example is impossible to replicate. However, heavily competitive environments act as a proxy for this example, since demand at a given casino is reduced by the presence of immense competition.

Pearson's Correlation

Pearson's Correlation measures the strength of dependence between two variables. The table in Figure 2 is a series of (x,y) scatter plots and the associated Pearson's Correlation coefficient of x and y for each set. Note that the correlations illustrated reflect the non-linearity and direction of a linear relationship (as in the top row), but not the slope of that relationship (middle row), nor many aspects of non-linear relationships (bottom row). Also note that the figure in the center has a slope of zero, but in that case the correlation coefficient is undefined because the variance of y is zero.[3]

The graph in Figure 1 illustrates a correlation of 0.66, which may look something like a merge of the second and third images in Figure 2. We can also see that a 0.8 correlation can, to a human, look like a clear dependence between two variables.

Figure 2: Correlation Examples

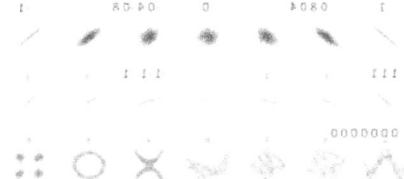

Yield Case Study

Penny Alley area was a low-capital initiative within Silverton Casino that resulted in a 19 percent increase in theoretical win per device.4 Our example here is based on the Seasons Buffet and its aptly named mini casino, Seasons.

Originally, the Seasons mini casino was the second largest mini casino in Silverton in terms of machine count. Seasons is adjacent to the buffet and has a high correlation (0.66) between buffet counts and player counts in the area. Before yield management alterations were made, 80 percent of the games underperformed, and although the area had high observed foot traffic, it had low retention, signaling that players did not find the area attractive. In Figure 3, the heat map diagram of the revenue of each slot machine shows that while there were some concentrated areas of revenue, the gaming floor before the changes was clearly a low-utilization area.

Figure 3: Seasons Mini Casino Before Change

Before making changes to Seasons, Silverton investigated whether the lack of attractiveness of the area was related to the product or to the layout, or both. Seasons mini casino contained a high number of unpopular, lower-denomination video poker machines and several banks of uprights laid out vertically through the center of the area, creating a visual wall. This led to the initiative to enhance the area by both corrections of the product and improvement in the layout.

The slots improvement initiative, lead by Salinda Conklin, VP of slots for Silverton Casino, had three main components:

1. Removing the visual barrier that split the section. This created a clear path to another gaming area and a restaurant that was previously hidden behind the barrier.
2. Cleaning up the game mix based on the product demand in the Seasons mini casino.
3. Adding more visually appealing slot configurations (e.g., several rounds, shorter banks).

Overall, the unit count was decreased by 27 percent. The remaining mix was tweaked to include more of the games that players preferred in the area and also games that players in that area tended to play elsewhere on the floor.

These changes to Seasons resulted in an increase in total revenue with fewer games, and therefore a dramatic increase in the yield. Total revenue for the area increased by 2 percent, despite removing over a quarter of the games. Yield (in this case measured patrons relative to machine count for the area compared to the floor) increased 25 percent. (See Figure 4)

Figure 4: Change Results

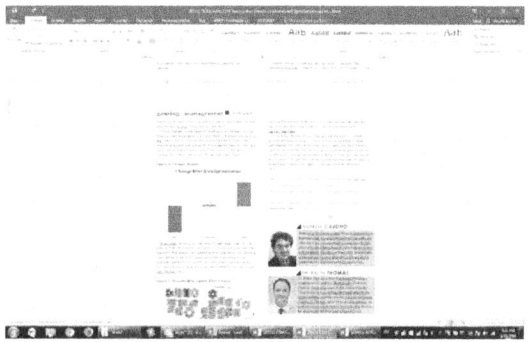

Now, when looking at the post-change heat map of the gaming floor of Seasons (see Figure 5), it becomes immediately apparent that players spread their play across the gaming floor. While there is still opportunity for improvement (for example, banks 05-037 and 05-022), this improvement can be executed in the context of a much more effective overall layout and product mix.

Figure 5: Seasons Mini Casino After Change

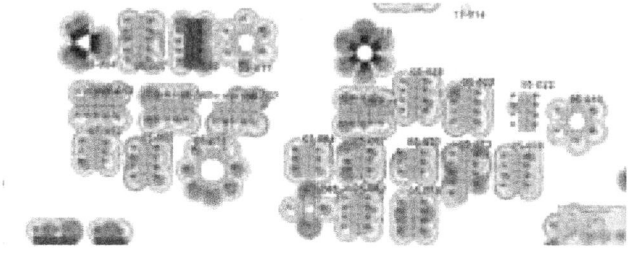

Finding the Money

So, in our Seasons example, we discovered a lower bound on the optimal spacing between games. Whatever the density was a year ago (call it t_0), we found that the optimal spacing t^* between games satisfies $t^* > t\_0$.

This is an important realization for casinos in highly competitive environments. This model can also be applied to any environment, albeit with very different results. In both a high-demand and a low-demand environment, the Laffer Curve of gaming density versus customer revenues will apply; however, the graph will look much different for high-demand versus low-demand environments. In particular, the optimal spacing t^* will be much lower (i.e., will provide for less space and more games) the higher the gaming demand.

These results show how the yield management of gaming machines can be a profitable exercise. Furthermore, they indicate that careful analysis of the density of gaming machines and the layout on the floor may result in profit improvement with less capital deployed.

From here, the key issues that we will explore are how to apply customer data to dig deeper, how cannibalization could affect our results, and how new metrics such corrected theoretical win can be applied to gain actionable insight across the whole property. What is truly exciting about these actions is that they often involve minimal capital, and yet, as we have shown once again, they can drive incremental revenue.

---

1 CEM, April 2010 "The Long and Short of It: Slot Games from a Player's Perspective," Singh, Cardno, Gewali.
2 http://en.wikipedia.org/wiki/Arthur_Laffer, extracted September 2011.
3 http://en.wikipedia.org/wiki/Pearson_product-moment_correlation_coefficient, extracted September 2011.
4 CEM, December2010-March 2011, "Mini Casinos Meet Mini Games at Penny Alley: A Case Study of Silverton Casino," Cardno, Singh, Thomas, Evans.

CHAPTER 6: PLAYER EXPERIENCE AND SLOT OPTIMIZATION

Authors' Note: This article part will define two kinds of metrics: optimization metrics and outcome metrics. We will also explore how metrics that are related to the player experience can be used as the central driver of gaming floor optimization. In future article parts, we will explore how we can use mini casinos combined with optimization metrics, such as corrected utilization and corrected theoretical win, to explore revenue optimization.

Ten years ago it could be argued that the best way to make money in a casino was to open one. In those good ol' days it was a reasonable optimization strategy to simply keep the best-performing machines, and optimization strategies based on this approach were quite successful. In today's world, however, many markets are saturated and now it is all about beating the competition, and, in many cases, this competition comes from non-traditional sources. For example, we now compete with online entertainment.

We have spent a great deal of time thinking about slot optimization and looking for methods to increase gaming revenue by making improvements to the slot floor. Exploring the philosophical side of gaming floor optimization, we can ask "What is slot optimization?" From the slot operator's perspective, the answers can include: Optimizing win per unit per day (WPU), replacing poor performers (based on WPU) with new product, and the effort to make sure each machine is driving its fair share of the budget (via WPU).

There exists a wealth of optimization tactics, including this slot optimization philosophy: "But how do we know when the gaming floor truly is optimized and we can go play golf? Well, when the incremental WPU for each machine is the same!" We believe that understanding incremental gains is fundamental to optimization efforts, however, we also believe that true slot optimization is best defined from the players' perspective.

The Players' Perspective

To truly understand gaming floor optimization, we have to put ourselves in the position of the players. Players do not, at least in our experience, make any decisions regarding the WPU of the gaming machine. And since the floor hold percentage is more dependent on the mix of gaming products than the hold of the gaming machines, players are also not interested in the overall hold percentage (although they are likely to respond to different pay tables on video poker). Finally, in the past, denomination (denom) was a key driver of player choice. But in this new era of penny games with a wide range of configuration options, denom is no longer the driver of player choice.

From this, player-centric optimization is defined as improving the player experience in ways that drive the most incremental player net revenue. Expansion of this definition results in two kinds of metrics: optimization metrics and outcome metrics.

- **Optimization metrics**: These are metrics that measure effects that players can observe. Furthermore, it is generally desirable to optimize these metrics to drive incremental revenue. In short, optimization metrics are metrics that players notice. Utilization is one metric that players notice—you might hear statements from patrons like "I found my favorite" or "the best games are always occupied when I want them."

- **Outcome metrics**: These are metrics that players do not observe. For example, the theoretical WPU on the gaming machine. It has been shown that players' experiences[2] differ dramatically from expected or theoretical outcomes, and furthermore, the theoretical WPU is an average from a number of different players. Quite simply, players do not experience the spending of other players. Another outcome metric is the slot floor hold percentage, or what is often incorrectly termed the "price" of our games. In the partour article "Hold Another Sacred Cow", we debunked the myth that one can increase gaming revenue by simply tinkering with the overall floor hold percentage.3 Quite simply, the hold percentage of the whole gaming floor is an outcome metric that is more reflective of the mix of gaming products than players' experiences.

Digging deeper into our player-centric slot optimization philosophy, the goal is to improve the player experience, thereby driving revenue and beating the competition. One cannot improve the player experience by watching WPU numbers. Rather, we must make changes to our slot floor to improve the player experience, and do so in ways that drive incremental player spend.

Consider this illustrative example: Imagine having an extremely popular slot machine. WPU for this slot machine is four times the floor average in WPU. Looking at some of the optimization options we encountered previously, it is tempting to look at that game and say "great, it's doing its job" and then focus on improving the WPU of lower-performing products. But now, looking at optimization from the players' perspective, a game with a WPU of four times the house average is a potential problem and, furthermore, a revenue opportunity.

If this high-performing game is in the high-limit room, it is likely to have low utilization. But if it is in a high-volume area of the floor, it might be a problem. Quite simply, it may have a negative effect on player experience. In an extreme example, if the game is 100 percent utilized, players are in effect always competing to play. Furthermore, these players may choose to play at a competing property where the product is more available. In this situation, the opportunity exists to optimize utilization to enhance the player experience and drive incremental revenue. As such, we can apply utilization to drive incremental revenue by enhancing the player experience.

Of course, that's all fine and good as a philosophy and for this simple example, but in the real world our patterns are finer grained, our data is large and many dimensioned, our players are numerous, and our capital is limited. The next section introduces some methods for handling these real-world challenges.

Slot Change Implementation Metrics

Let's take this to the next level by looking at optimization metrics in more detail and showing how statistical clustering methods can be applied to this data to provide real-world optimization.

The player cares about time on device and their gaming experience. Put another way, factors that drive how much a player expects to lose over an hour of play (SPH) include:

- **Average Bet** – Certainly how much the player spends per play is a big part of SPH. The average bet itself is a complicated function of the configuration of the game via denom, minimum bet, maximum bet, maximum bet to cover all lines, etc.

- **Game Speed** – A slow game can provide for longer time on device, whereas a fast game can provide for more exhilarating action. The player can choose which speed suits their play style better, but this factor is vital to understanding game performance in general and SPH in particular.

Once we calculate SPH, we can better understand the "cost" of each of our games from the players' perspective of time on device.

So now our player has selected a game that provides an SPH that he/she is comfortable with. The next big question from the player is "Where is my game?" Studying the location of games is a critical part of this.

Figure 1

	Low Spend Per Hour Player	High Spend Per Hour Player
Low Spend Per Hour Machine	Gamer	Grinder
High Spend Per Hour Machine	Loser	Gambler

By cross tabulating the quadrants of SPH of the player versus SPH of the machine (and leaving aside the debate of whether we should be using median or mean or Monte Carlo simulations), four categories of playing experience are created (see Figure 1):

Grinders: These players are spending below their typical SPH. If this is because they cannot find the gambling experience they want, then we have an opportunity.

Losers: These players are being hammered. Unless they are lucky, they are unlikely to have the wallet to continue this gaming experience.

Gamers: These players are spending at low amounts and often represent the majority of the utilization on the gaming floor.

Gamblers: These players probably drive the majority of the revenue in the property. It is critical that we optimize their gaming experience to ensure that they find the gaming devices they want.

Quite simply, we want patrons to have the gaming experience they are looking for. This experience should be neither above nor below their "expectations" and should be on the game they desire at the location they desire.

Applying these player experience concepts to groups of players or individual players is even more interesting. For example, a high-value player may be grinding for a while but then go back to gambling.

When building the optimization models around this data, it is critical that we consider these gaming experiences. To illustrate this point, consider what would happen if the optimization model mixed these four player experiences into one number. This WPU number would probably be driven by gamblers but the machines would likely be occupied by gamers. One model at least one of us authors has applied is to maximize gambling and minimize gaming.

Another question that players ask is, "Is my game available?" To answer this question, we leverage utilization. The calculation is simple: What percent of the time that a game is available is it being used? Utilization is often calculated on a 24-hour basis, meaning that whether a game is being played at 4 a.m. on a Tuesday or 7p.m. on a Friday, the amount that play contributes to the overall utilization score is the same. To handle this, consider utilization as one method to simply compare games. If one game has 50 percent utilization and another has 10 percent utilization, it is far more likely that the game with 50 percent utilization is going to be busy when a player wants to play it than the 10 percent game.

Let's finish this particular discussion with another illustrative example. Suppose there are three games available and we need to decide which game needs more units. In terms of our metrics, we have:

A. 50% utilization and $15 SPH
B. 20% utilization and $50 SPH
C. 5% utilization and $250 SPH

Figure 2

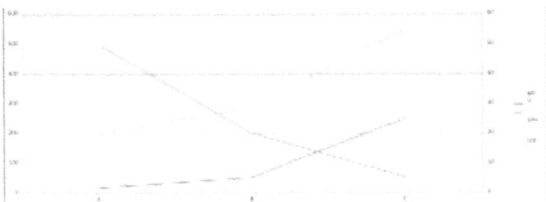

Let's calculate WPU for each scenario, taking the utilization x 24 hours in a day x the spend per hour. The WPU for Game A is $180, Game B is $240 and Game C is $300. A traditional WPU analysis of these games would leave an analyst thinking that Game A is the weakest and that no more units are required. However, what if it was discovered that incremental spend from the player occurs once utilization crosses a certain threshold, say 30 percent? If that were the case (and significant slot change analytics needs to be done to determine this threshold), then only Game A could provide incremental spend via incremental units. Adding more of Games B and C may only result in diluting the performance of the existing games. (See Figure 2.)

Utilization

So far we have left utilization as a generic concept. In practice, overall utilization can potentially be as useless as overall floor hold percentage. A player who only plays on Friday nights does not care about the overall utilization of a game; they care about the utilization of a game on Friday nights! As was mentioned above, this may not be an issue in less competitive environments, but as our industry gets more and more competition, and as our supply and demand reach close to equilibrium (or even oversupply), understanding utilization at the proper time periods becomes more and more important. However, once we place a game, we are stuck with it 24 hours a day, until we choose to replace it, though server-based games have long held the promise of helping with this time of day optimization. One powerful method for handling the dimensionality of optimization across different time periods is clustering.

Clustering

In the past it could be said that operators have assumed that they have one player. Player-independent metrics like WPU and floor hold percentage were used, and the operators might say things like, "Don't worry about where you place that game, the player will find it!" Lost in this view is the fact that we have thousands upon thousands of players. And while these players may have behaviors that we can average down to statements like "the player spends $100 per visit," this actually applies to very few of our players. As an exercise, take the average spend of your players, then see what percent of your players actually have an average individual spend within 5 percent of the overall average. Most likely the vast majority of your players do not actually fall within this range.4

Using methods discussed in "Clustering: The Key to Understanding High Dimensional Data,"5 group your players by not only the games they choose to play, but also the time periods they choose to play in. By adding this dimension of time to our clustering methodology, the analysis measures utilization across various player clusters and gives us a better understanding of which games cause our players to answer the question, "Is my game available?" with a frustrated "No!" We can relieve that frustration by getting more of that game, thus improving the player experience.

The concept of player clustering can be a difficult one to grasp, so here is a clarifying example. Instead of the complicated question of, "Is the game available for the player when they want to play?" we ask a simpler question: "Do players make their game decisions based on the look of the game or on the play mechanics?" At least one of the authors performed a simple clustering analysis of slot product and discovered that the answer to the question is both.

There are certain player clusters for which play mechanics is the primary driver of their game choice. We see this when manufacturers offer "clones" of their games, meaning the math and game rules are the same but the symbols and pictures on the cabinet are different. Certain player segments have identified that this play mechanic is the one for them, and they play all different versions of this product, regardless of how the game looks.

However, there is a completely different cluster of players that are attracted to what can only be described as "exotic" games. These are the games we often see at industry conventions, with bright lights, fancy seats and, often, pop-culture references. These exotic game chasers love the look of a new game and could care less that these games come with a wide variety of different play mechanics.

Conclusion

This non-WPU method for slot floor optimization sets out to enhance the player experience and use this enhanced experience to beat the competition. We have been applying these methods with some remarkable success, and as our philosophy of gaming analytics advances, we expect to see further refinements in how to drive this incremental revenue. Of course, the choice is yours. You can continue to optimize using WPU and hope that the methods that worked so well 10 years ago will continue to be effective, or you can switch to a player-centric, and likely more profitable, approach to optimization.

1. "An Analyst's Guide to Slot Floor Optimization," November 2010, Casino Enterprise Management.
2. "The Long and Short of It: Slot Games from a Player's Perspective," Singh, Cardno, Gewali, April2010, Casino Enterprise Management.
3. "Hold Another Sacred Cow," Cardno, Thomas, April 2011, Casino Enterprise Management.
4. "Clustering to Uncover Hidden Behavior: A Case Study of Silverton Casino," Cardno, Thomas and Evans, April 2011, Casino Enterprise Management.
5. "Clustering: The Key to Understanding High Dimensional Data," Cardno, Singh and Thomas, February 2011, Casino Enterprise Management.

CHAPTER 7: FINDING THE MONEY IN JACKPOT WHARF, PART 1

Authors' Note: This is part 7 of our 18-part series on "Where's the Money." This is part 1 of a two-part series on Jackpot Wharf. The second part will look into customer displacement and scientific control. This article part digs deeper into the measurement of success of gaming mix optimization in the Jackpot Wharf area of the Silverton Casino in Las Vegas. The objective of this project was to create a destination-style gaming area the builds on the high level of non-local customers in the property. From a very high level, the task was to identify games that had high levels of non-local play and cluster them together around the mermaid tank. This analysis applies metrics such as spend per hour (SPH), utilization and percentage of non-local while considering yield per square foot. One important attribute of the analysis was taking into account what was a significant reduction in the number of gaming devices.

Silverton Casino is an establishment that generally caters to locals. However, it has a unique source of tourists—a Bass Pro Shops store that is literally connected to the casino. Heavy foot traffic can be observed between Bass Pro and one area of the casino (Jackpot Wharf) that houses a 110,000-gallon salt water aquarium that occasionally features live mermaids. Jackpot Wharf is situated on the eastern end of Silverton Casino, which has long been a low-performing area of the casino.

Silverton Casino launched an initiative to transform Jackpot Wharf into an area that would convert the mostly tourist foot traffic visiting the aquarium into gamers. This mini casino is the third step in the ongoing optimization of the gaming floor; (The first step was described in the CEM series on Penny Alley, and the second was the recent Seasons optimization described in the November 2011 issue).

Mini Casino Strategy

To implement this type of mini casino optimization strategy, there are a number of critical requirements. Without these requirements, it is quite simply not practical to implement a mini casino strategy.

1. **Naming of areas.** While this seems like a simple step, it is in fact one of the most difficult tasks in building a strong mini casino strategy. The name is key to unifying the slot management and marketing groups. Unified operations and marketing are key elements of the success of the mini casino strategy. In implementing this strategy, it is important to recognize that it is not just slots or patrons alone that generate revenue...it is patrons spending money on slots.

2. **Spatial management of physical spaces.** The definition of physical spaces is a traditional mapping problem. The challenge is not the initial definition, which is what at least one of us authors thinks is a relatively straightforward process; rather, the challenge is the ongoing management of the relationship. Consider moving slot stands on the gaming floor. What is needed is a system where the slot stand moves (not to be confused with asset moves, where no further information is required), automatically updating the mini casino that they belong to based on the spatial relationship.

3. **Locational intelligence and spatial data query.** As people use locationally-enabled mobile devices to do things such as make restaurant reservations or find the nearest bathroom, this data becomes interaction data, showing how customers interact with us in our business. It is reasonable to contemplate this interaction data being many thousands as times as large as the transaction data we currently analyze. In the brick-and-mortar business of casinos, understanding

where these interactions take place requires spatial query, and mini casino regions are a key part of this.

4. **Seamless integration into relational data using implicit data relationships.** The modern relational database seamlessly blends relational and spatial data. It is now possible to write queries such as "show me every patron who has played within 5 feet of a host who did not give them a drink" or "show me all the high card tier players who turn left when they enter the main door." This seamless integration of both spatial and relational data into one database opens some analytical doors that used to reside firmly in the domain of a select few specialist tools.

Silverton Casino's War Room

The War Room at Silverton Casino is a central location for its observation process. From time to time, the whole team gathers in this room and collaborates to define new initiatives and to assess the results of current initiatives. The War Room consists of a large number of plotter-sized printouts covered in hand-written notes and letter-sized graphs of results. The goal is to bring the analytics into central focus, with all the numbers on the wall and all the team members focused on creativity-finding initiatives.

In addition to its collaborative benefits, the War Room provides the ability to leverage a new kind of analytics: mega-sized data visualization. In the past, we were restricted to 8.5-inch x 11-inch sheets of paper pressed together in a binder. With mega-sized data visualization, we can essentially "zoom out" and see larger chunks of analytics in a single view, allowing for observations across the data that normally would not be seen. Instead of flipping and searching through that binder, we stand and browse the data across the walls of the War Room.

Observations and Strategies

The initial retention rate in Jackpot Wharf was very low, and financial performance was mediocre at best. The silver lining was that rated, non-local financial performance was off the chart.

Observations of the play patterns in Jackpot Wharf indicate that this area tended to have a very high level of non-local players and a high level of non-carded play. In examining the data further, we noticed that Strip-style games were heavily played and that the locals-style gaming products were under-performers.

The low retention rate indicated that customers were dissatisfied with the product, the layout, or both. The fact that we saw high non-local performance further indicated we had a product imbalance. The first step to balance the product was to find out who was playing what product, where else they were playing, and how similar products were faring elsewhere on the floor.

We found that the non-locals were primarily playing the participation games in Jackpot Wharf, followed by other small segments of house product. Locals who played in Jackpot Wharf spent a small part of their overall time there, preferring other areas of the floor. Understandably, product that was suffering in Jackpot Wharf was generally product that locals preferred and that was thriving in other areas.

The second issue we wanted to address was layout. We wanted the new participation games in prime locations—essentially showcasing them—with the intent of attracting as many impulse purchases as possible. We improved the grocery-aisle style layout by introducing several rounds. This was especially effective by the front doors. Where patrons used to immediately walk into two banks of machines, they are now drawn into the area through the use of a shaped bank, which has improved traffic flow to the games along the wall.

The Mini Casino

The character of a mini casino is critical and is a key part of the final result. To quote Barry Thalden: "Character of the casino floor is a huge factor in casino performance. We have seen same-slot revenue increased by 30 percent to 300 percent in areas that are intelligently redesigned."1 This leads to further opportunity for architectural innovation where the character of the mini casino is further enhanced to drive significant revenue. In our CEM series on horizontal innovation, we defined horizontal innovation as "innovative technologies on the gaming floor that apply broadly across multiple gaming devices", in the September 2011 issue.

Character vs. Function

Let's consider the fundamental difference between function and character through consideration of a vehicle. The function of a vehicle is "any means in or by which someone travels or something is carried or conveyed."2 This functional description of a vehicle defines what the vehicle must do to be classified as a vehicle. However, the character of the vehicle can be anything from a spaceship3 to a motor vehicle. Clearly, the characteristics of the vehicle are extremely important in defining what it is; otherwise, we would confuse spaceships with motor vehicles.

Character of Jackpot Wharf

In the world of mini casinos, all mini casinos are functionally similar, if not identical. It is the character of how this function is delivered that is important. The character of Jackpot Wharf was driven by its customer base: a little taste of the Las Vegas Strip with games that are not popular with locals, including heavily themed games such as *The Hangover* and *Sex and the City*.

The Results

While the final results are yet to be determined after a longer period of analysis, the initial results look very promising. Figure 1 shows how the net revenue has increased substantially. Overall, the results show an approximately 25 percent increase in net revenue for the mini casino. What is more remarkable is that the casino overall has seen a small increase in net revenue in the same time frame. This increase poses questions, such as if the character of the Jackpot Wharf mini casino increased the overall performance of the property.

Figure 1: Net Revenue Increase

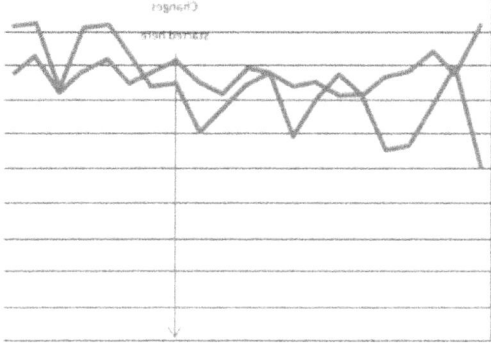

NOTE: Red is floor; blue is Jackpot Wharf. Tick marks on the X axis indicate weeks.

Jackpot Wharf

Looking at Jackpot Wharf itself, we see a dramatic upward swing in all metrics in the area. (See Figure 2.)

Figure 2: Jackpot Wharf Performance Trend5

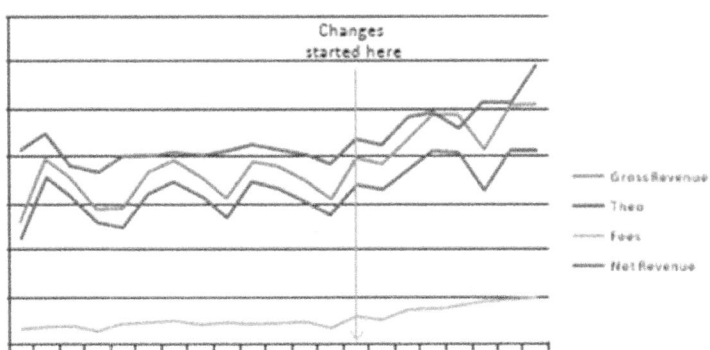

Tick marks on the X axis indicate weeks.

Conclusion

The results from Jackpot Wharf have, at least so far, been nothing short of spectacular. Clearly it is a further validation of the mini casino strategy. Furthermore, from Silverton Casino's perspective, it shows how the casino can design an initiative that focuses on the unique competitive advantages of the property. The Jackpot Wharf initiative shows how we can cater to different market segments in one property and that cannibalization from a mini casino focused on one market, for example Penny Alley, to a second mini casino, Jackpot Wharf, can be—and in this case was—quite minimal.

Each step of advancing the Silverton property has resulted in a refinement in its mini casino initiatives. What is truly exciting is to see how the marketing and slots groups are cooperating so closely and how this cooperation is delivering bottom line results.

With the continued success of these revenue-increasing yet low-capital initiatives, it is the choice of the operator to apply mini casino-based optimization—a choice that may be hard to say no to in many markets.

1. Barry Thalden, quoted with permission November 2011.
2. Extracted from http://dictionary.reference.com/browse/vehicle December 2011.
3. At least one of the authors loves space ships.
4. The axis of this graph has been left of deliberately to obfuscate the data.
5. The axis has been deliberately left off to obfuscate the data.

CHAPTER 8: PLAYER PREFERENCES LEARNED FROM JACKPOT WHARF, PART 2

This is the second article part of the "Where's the Money?" series that focuses on Jackpot Wharf, a mini casino area within Silverton Casino. This article part will look into player displacement and clustering analysis of the player behavior. It will also introduce a new concept, preference filters, which are designed to highlight areas of preference for gaming product on the gaming floor. Preference filters are a critical tool for discovering what drives play from players who have a preference while filtering out players who spread their play widely across different gaming products.

In our last article part, we covered how "war room" collaboration drove an ambitious effort to transform Jackpot Wharf from one of the weakest areas of Silverton Casino to one of the strongest. The overall goal was to have the Jackpot Wharf area offer a taste of the Las Vegas Strip experience, enticing the destination market customers who visit the property for its Bass Pro Shops store to stay and play. The changes to the area drove the revenue up by more than 20 percent, and now Jackpot Wharf is the strongest performing area of the gaming floor. Furthermore, the additional revenue has been accompanied by an increase in year-over-year revenue on the balance of the gaming floor.

As we dig deeper into the results of our Silverton Casino analysis, our core goals are to enable learning from the successes we find. There is now an expectation that, given the success of Jackpot Wharf and the earlier success of Penny Alley (another mini casino at Silverton Casino - see the December – March 2011 issues of CEM for the complete analysis), further work with mini casinos will drive incremental revenue. However, learning from one's successes is sometimes harder than learning from one's failures.

According to Robert Sutton on his Harvard Business Review blog (June 4, 2007), "After people succeed at something, it is especially important to have them focus on what things went wrong. They learn more than if they just focus on success (so, don't just gloat and congratulate yourself about what you did right; focus on what could go even better next time)." In other words, if we can find the cause of any missteps on the road to our success, as well as identify the key drivers of the success, we will be able to replicate that success next time...and maybe even improve it.

Because we're dealing with a mini casino, understanding player preference is an extremely important part of the analysis.

Preference Restrictions

A preference restriction is a means of grouping players based on the manner in which they select their gaming product. Here are some examples of preference restrictions:

1. Select all players who played on Machine No. 123 in December.
2. Select all players who played on Machine No. 123 in December and who also had at least 50 percent of their total play appear on Machine No. 123 during that month.
3. Select all players who have at least one machine that captured 50 percent of their total play in December.

We can also request additional data from these restrictions in multiple ways. For example, for Preference Restriction 1 above, we can choose to look at those players' overall play data in two very different ways:

- For all players who played on Machine No. 123 in December, select their play across all other machines during that month.
- For all players who played on Machine No. 123 in December, select their play across all other machines during that

month, but only on days that they also played on Machine No. 123.

In the above examples, we selected groups of players who passed our preference restriction during December, but we could have easily selected a specific week in December or even a single day. However, this gets much more complicated when we consider the fact that players behave differently on different gaming visits. As a simple example, what do we do with a player who only plays Machine No. 123 on December 1, but who then returns on December 15 and ignores Machine No. 123 entirely in favor of Machine No. 456? Do we conclude that this player is 50-50 about these two games, or was the player just tired of No. 123 and tried No. 456 on a lark? This is not an easy question to answer, but we need flexibility in our data to allow for the possibility to research these types of questions.

To do this, we need to look what we call the "player visit." Instead of looking at the play of each player over the month of December, we go down a level deeper and look at the play of each player on each visit to the casino, and consider these separately. So our player's visit on December 1 is considered completely separately from his visit on December 15. In our example above, when looking at Machine No. 123 and our player who visited on the 1st and 15th, we would only consider the player visit that occurred on December 1, since that is the visit when the player played on Machine No. 123.The play from his visit on December 15 would be completely ignored in the analysis. In practice, it is the experience of at least one of the authors that, when using preference restrictions, the information that is derived when analyzing player visits is far more impactful than that derived when analyzing players only.

Preference Filters

Before we dig deeper into discovery of the gaming floor revenue, let's look an interesting algorithm called preference filters. To understand preference filters, think about a casino with two players and ten slot machines. These players are categorized as Play

Everywhere (PE) or Play Two (P2). PEs play on every game, and P2s play on only two games.

Figure 1 shows a graph of the revenue per day for these two players. A look at the gaming floor shows that Location 4 is the strongest, followed by Locations 3, 5, and 6. The underlying question is: are there location preferences amongst these locations?

Figure 1: All PE and P2 Data

To answer the question of preference, let's filter out all players where less than 16 percent of their play is at that location (see Figure 2). (Note: This 16 percent of play means that the ratings for players will be filtered to, at most, the top six locations and, in practice, only the top two or three.)

Figure 2: Theme Preferences

Now the results look quite different. We can see that Locations 3, 4, 5, 6, and 7 are driven by play from players who have a strong preference for these locations. This example illustrates the concept of preference filters, and in this case, the numbers are easily understandable. When applied to the larger context of real gaming floor numbers, it still serves to highlight locations that players prefer to play at—a powerful and simple concept. Preference filters become even more interesting when they are calculated against different aspects of the data.

The most important characteristic of a preference filter is that the calculation of the filter is executed at the session or ratings level. This low-level filter means that each view of the data is a new calculation of the results. The challenge with these preference calculations is that they truly exploit detailed data, and that data is often very large. This means that the methods of building aggregates or data cubes are not useful for preference filter calculations.

Now let's examine some more sophisticated preference calculations.

Question 1
What is the First Choice in Game Theme?

The masses of players moving around all have very different game preferences, but in our experience, these groups of players often fall into distinct groups of players. (See "Market Basket Analysis," CEM, December 2008). Some players are prone to play on many different themes, while others concentrate their play on only specific games. This combination of different play patterns creates a kind of noise, where many groups of players play across a wide range of gaming product.

By filtering the play to only players who have a preference for a theme, the concentration of play on the gaming floor now shows the magnet games or product that players will almost always play. For example, if we apply a theme-based preference filter to only show players who spend at least 50 percent of their time on one specific theme (regardless of where that theme is), then the resulting analysis is going to be highly concentrated on games that are magnet or first choice games themes. This preference filter can be further refined using subsets of players. For example, "Show me only local players versus destination players."

In the case of Jackpot Wharf, the game theme preference filters showed a strong tendency for local players to not play in the Jackpot Wharf area. Furthermore, the preference filters showed that players who played in the non-Jackpot Wharf areas of the gaming floor did not move their play to the Jackpot Wharf area.

The theme preference filter was very powerful in understanding this pattern, as we were able to see that the local players remained with the same preferences and that they did not change to play on any of the new product located in the Jackpot Wharf mini casino.

Question 2
What is the Displacement of Players with a Preference?

The alteration of games on the casino floor and the subsequent displacement of play patterns is one of the critical questions in gaming. (Note: We have intentionally obfuscated this example to protect Silverton's confidential data.) Using the preference filter, we can calculate how many players prefer to play on a bank of slots that we are considering changing.

From the preference calculation comes a powerful selection process. Looking at Figure 3, we can select the players who preferred to play on Bank 04-007.This bank is an illustrative example of a bank that was changed and how the preference calculator can be applied to understand the displacement of game play. Clearly the players on Bank 04-007 are concentrated on these banks and are not playing on some of the surrounding games.

This displacement of gaming revenue is one of the core questions that allows for ongoing optimization of the gaming floor. In addition, the selection of players with a preference enables the marketing team to determine who stopped visiting the property. Combined with a change in gaming product, this change in visitation may well have resulted from the alteration of a product that those players preferred.

Figure 4 shows the same group of players who preferred the old games at Bank 04-007 and how they altered their play patterns after the game had changed. Notice that this group of players did play on the new 04-007 games, but, at least in this example, the players who preferred the old 04-007 game were displaced to other banks, including 04-064 and 04-061.This examples illustrates how a preference filter can be applied to select a group of players to further examine the displacement of their gaming revenue following a change in the gaming product.

Figure 3: Bank 04-007 Players

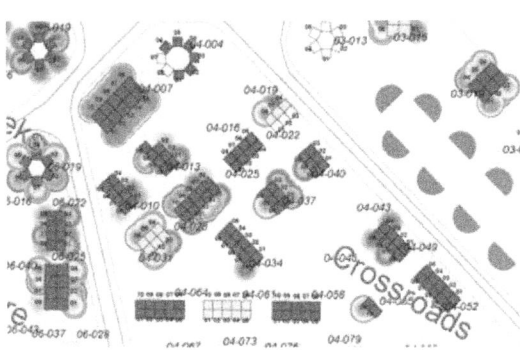

Question 3
How Do Math Models Affect Preference?

Math models are one of the core features of a gaming machine, and many themes share the same paytable. The preference calculation based on the preference for a gaming model enables the analyst to address questions such as, "Are there gaming models that are driving magnet products?" Using the selection process, the players with a preference for a paytable can be selected and further analyzed to see if there are other models that they will play on or what kind of volatility of gaming model that they enjoy.

Figure 4: Bank 04-007 Players After Game Change

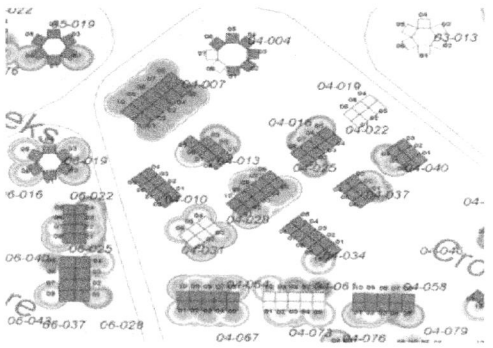

More Results from Jackpot Wharf

Jackpot Wharf truly achieved its goal at Silverton Casino. During November, the marketing department executed an aggressive hotel campaign that resulted in a 150 percent increase in gaming business from hotel guests. Furthermore, as desired, the Jackpot Wharf mini casino responded with a more than 250 percent increase in gaming revenue from this marketing program in the new targeted area.

The increase in non-rated play by over 33 percent in Jackpot Wharf has been a further reinforcement of the targeting of this area to the non-locals market (local players tend to be carded). Given the volume of the changes, it is the belief of the authors that the redesign and refocus of the Jackpot Wharf mini casino was the central causal factor in these increases.

The bottom line is that Jackpot Wharf continues to have very little impact on locals play and has driven the play from non-locals and non-rated players.

Conclusion

The calculation of game preference is one of the most powerful techniques in deciphering game play patterns. The case study at Silverton shows that the combination of the mini casino strategy and the preference calculations can, if executed correctly, drive significant revenue. Numbers like a 250 percent increase in revenue from a targeted group are hard to argue with and stand as further validation of the importance of insight-driven business initiatives. Operators now have a choice whether to adopt these powerful techniques that enable them to filter out the noise in the data and see true player preference and displacement or to continue to optimize the gaming floor based solely on revenue metrics.

CHAPTER 9: BIG DATA

In this article part, we will look at "big data," what it means for the gaming industry, and specifically, what it means for the gaming floor. Big data is essentially ultra-large datasets that are being updated frequently. These datasets are at the forefront of many of the biggest changes in the world today—Google, Facebook and Yahoo! are three canonical business at the epicenter of the big data explosion. The gaming industry has a deep history of exploiting data, and big data is unlikely to be the exception. This article part will start to prepare you for the changes that big data might bring.

Data Then and Now

The last century's data breakthrough was transaction data. In other words, the finest-grained data was collected by retailers tracking their sales and inventories. Transaction data was considered to be massive, and a highly specialized industry evolved to handle these huge data volumes. The shining example is Walmart and its 2.5 petabyte database that continues to drive its global retail link solution. [1]

This century's super star data, however, is interaction data. The graph in Figure 1, based on extrapolated International Data Corp. (IDC) estimates, shows how we are now crossing into the Zettabyte Age. In this data age, the majority of information comes from interactions with customers. The key attribute of interaction data is that it happens before and after transactions. Quite simply, this means that the potential exists to see what customers are shopping for as well as what they actually choose. Already there are companies whose main value driver is using and exploiting this data. You do not have to look further than Google, Yahoo!, LinkedIn, and Facebook to see companies who are pioneers in this developing space.

Figure 1

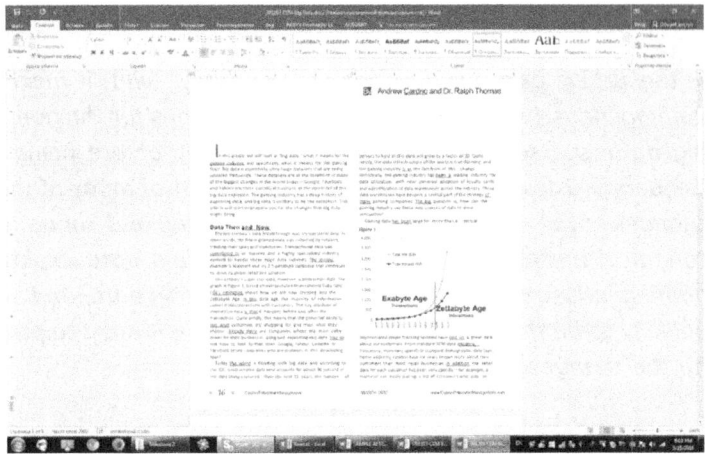

Today, the world is flooding with big data, and according to the IDC, unstructured data now accounts for about 90 percent of the data being captured.2 Over the next 10 years, the number of servers used to hold all of this data will grow by a factor of ten. Quite simply, the data infrastructure of the world is transforming.

The gaming industry is at the forefront of this change. Historically, the gaming industry has been a leading industry for data utilization, with near universal adoption of loyalty cards and a proliferation of data warehouse enterprise data assets (EDA)s across the industry. These data warehouse EDAs have become a central part of the strategy of many gaming companies. The big question is: how can the gaming industry use these new sources of data to drive innovation?

Sophisticated player tracking systems have told us a great deal about our customers. From standard RFM (recency, frequency, monetary spend) data to standard demographic data (e.g., age, home address), casinos have for years known more about their customers than most retail businesses. Furthermore, detailed data for each customer has been very specific. For example, a marketer can easily pull up a list of customers who play on Tuesday mornings with an average bet of more than $1.

More recently, companies that have installed enterprise-wide data warehouses (EDWs)data assets have been able to add a second set of dimensions to their data—product selection. By combining data from player tracking systems with data from slot accounting systems, casinos are now able to know every game played by every customer at every point in time.

Let's see what this does to the size of the data, assuming our casino has 500,000 rated customers in its database and 2,000 slot machines on its floor. For the sake of this example, we'll assume that each customer visits four times per year and plays an average of three different games per visit.

Traditional Player Tracking Data

500,000 customers x 4 visits per year = 2,000,000 records per year

EDW Data

500,000 customers x 4 visits per year x 3 slot machines per visit = 6,000,000 records per year

Dimensionality and Complexity

In a certain sense, the dimensionality of the data tells us how complex or "big" our data is. Dimensionality is relative to the problems that we are trying to solve. We could arbitrarily create big data sets if we wanted to. For example, imagine trying to cross-reference all of your customers against all possible sets of three-letter acronyms in the English alphabet. There are 17,576 possible three-letter acronyms, and if we were suitably crazy, we could create a data set that contains a last name column and one column for each three-letter acronym. Each row would then contain the customer's last name plus a 0 or a 1, depending on whether the three-letter acronym occurred within the customer's last name. This is a massively complex and large data set. It is also completely useless.

This example does, however, show how attribution of data can create more complexity than the numbers, and in a world of big data, there tend to be a lot of dimensions.

Let's take another look at the complexity of our data from the standpoint of dimensionality, comparing traditional player tracking data to the newer EDW data sets. For this example, we assume that we are interested in looking at the play of our customers over a fixed time period (1 month) and that there are ten reasonably interesting gaming metrics to examine for this time period.

Traditional Player Tracking Data Dimensionality

500,000 players x 10 gaming metrics = 5,000,000 records

EDW Data Dimensionality

500,000 players x 10 gaming metrics x 2,000 slot machines = 10,000,000,000 records

With the addition of EDW, our data explodes from a matrix of 10 columns and 500,000 rows to one with 20,000 columns and 500,000 rows—a dimensionality of 10 billion!
From the CEM article part "The Petabyte Era of Gaming Data," (Singh and Cardno, September 2008), we know that the move to detailed interaction data on the gaming device leads to a 480 times increase in the volume of data collected. Just 10 years ago, nearly every piece of data was touched or viewed by a human; today we only view 1 percent of the data generated. In another decade, humans will only see a tiny portion, probably less than 0.0001 percent of all data generated.

The last century's data was a sample of what was occurring in the real world. These samples of data, while still of considerable size, encouraged models that understood the nature of the data. In the future, when data volumes will essentially represent the entirety of all information, the challenge will not be building a model from a sample, but building a sample of the whole.

In addition, historically, the economic value of data has been gathered through careful analysis of transaction data. The operator controls the data, and the operator therefore controls the data's power. In today's world, "competing on analytics"3 is a well-established practice. This establishes analysis of transaction data as a requirement for business, and it is likely that this will expand to competing using the big data of the future.

Consider someone browsing online for a hotel to book—and, remember, with interaction data, we are able to see the customers' actions during their search, not just at the time of booking. The hotel operator that is able to respond to this shopping event is going to be acting before their transaction-based analytics have even started. In this sense, the interaction data we are now able to collect is like a crystal ball that can show us what our customers are about to do before they transact with us.

Data Advantaged Customers

Customers are, in a way, also early adopters of this interaction data, as both its creators and as users of how it can be applied. This information places the power of yield management in the hands of the customer. Consider a world where your customers have near perfect insight into your competition's offers and comparative views between them—your customers know the marketing programs and the responses from the market. In this world, the customer has perfect information and operators are at an information disadvantage. Let's call these customers "data advantaged."

Data advantaged customers already exist today, and it seems reasonable to assume that the number of data advantaged customers will grow over time. This growth is more horizontal in nature, as social media permeates different aspects of society—Facebook has 845 million users, 161 million of whom are in the U.S.[4]

These customers are also more communicative online than ever. To use Facebook as an example again, it records 2.7 billion "likes" and comments daily.5 There are companies now betting their futures on the likelihood that social media will replace e-mail.6 If these companies are correct, then not only will our traditional relationship with customers be fundamentally changed, but our method of communication will also be "shared" beyond our control by customers using social media.

If these changes do occur, then there will be a massive shift in information power toward the customer. In this environment, operators need to think about how to effectively take part in these interactions to try to gain back some of their information power. Two initiatives that organizations can follow to balance the power are locational intelligence and predictive analysis, each described below.

Locational Intelligence

Locational data, collected via GPS, proliferates in the world of big data; it accumulates on Facebook and through many smartphone applications. The locational view is extremely powerful in deciphering the mass of communication, and we can now understand customer behavior and target rewards that are appropriate, not only to what we think customers need or want, but more specifically, we can target rewards based on what customers needs or wants are at specific GPS locations.

GPS data aggregates to a location, and so vast amounts of location-specific data can be seen, understood, and acted on. As an illustrative example, we could display all the patrons who look for fast food while inside the casino and see if there is a relationship between where patrons are playing and their desire for fast food. We could, for example, then show the conversion rates to the in-property Johnny Rockets. If the numbers are low, maybe something as simple as better signage could keep the patrons on site for their fast food desires.

Predictive Analysis

Amongst the massive amounts of big data, there are deep trends and deeper analytics. There are companies driving this data to predict the future. For example, we can apply social media data to predict future hotel occupancy levels or to predict the number of attendees at an upcoming convention. This data has the potential to predict our upcoming business and to drive the interaction-based hotel yield management systems of the future.

Where is the Money?

The ocean of information that is interaction data only takes us to general trends. It is the transaction data that we hold that shows the conversion to real value. So by combining transaction data with the interaction data, we gain a chance of understanding both the behavior of our customers and the results of our actions. This transaction data is still growing at a tremendous rate and the integration with the interaction data is, at best, an unexplored space.

1. Information from www.informationweek.com/news/software/info_management/228800661 January 2012.
2. Extracted from www.emc.com/collateral/demos/microsites/emc-digital-universe-2011/index.htm January 2012.
3. Thomas H. Davenport and Jeanne G. Harris, Competing on Analytics: The New Science of Winning.
4. Extracted from www.itproportal.com/2012/02/02/facebook-releases-uvsage-statistics-845-million-users- 27-billion-daily-likes-and-comments/ January 2012.
5. Extracted from www.itproportal.com/2012/02/02/facebook-releases-usage-statistics-845-million-users- 27-billion-daily-likes-and-comments/ January 2012.
6. Reference www.telegraph.co.uk/finance/newsbysector/mediatechnologyandtelecoms/digital-media/8799510/Edward-Saatchis-private-social-network-aims-to-make-businesses-more-democratic.html January 2012.

CHAPTER 10: BIG DATA AND LOCATIONAL INTELLIGENCE

Authors' Note: This is the second article part in a subseries to investigate "Big Data." Locational intelligence is the data that joins brick-and-mortar businesses to social media-centered interaction data explosion. For an introduction to Big Data, please see the March 2012 issue of CEM.

The entertainment industry is firmly in the business of selling experiences—hopefully, exciting experiences. It is simply wrong to compare this to the retail industry, which is in the business of selling things. Of course, when shoppers shop in a retail outlet, they gain an experience, but the main goal of shopping is to arrive home with one's purchases. Entertainment is quite simply different. The main goal is to have an experience, and this is not something that somebody else can do for you—it is a personal experience. This fundamental difference between the industries also results in a fundamental difference in how we can use the data we collect about our customers.

Monitoring Twitter and Facebook1 has shown that there is a huge volume of information being generated by our customers regarding their gaming experiences. In addition, there is an even more significant volume of information being circulated by our customers as they react to other entertainment experiences. Looking at the volume of communication and its seemingly random nature can seem overwhelming, even more so when we consider that people communicate about entertainment at volumes that probably swamp all other aspects of their online communication. This communication largely takes place via social media websites and is about people sharing experiences. In many ways, the entertainment industry is at the center of an information storm, and it seems reasonable to speculate that a huge portion of social network interactions relate to the entertainment industry. For example, with a single YouTube video, a singer like Susan Boyle, a person previously completely unknown to the world, can become an overnight sensation and the center of global discussion.

So how do we start to sift through all of this data and use it to drive business? One of the most important, and easily actionable, data streams for casinos today and into the future is locational data. Let's take a look at how this data can be used in conjunction with interaction data and transaction data to unlock the value in the Big Data.

Locational Intelligence

Locational data was once strictly the domain of land surveyors and geographers. Today, it is generated by nearly every application on any mobile device. Consider this example: A customer named Andrew is sitting at a slot machine in his favorite casino, Casino Y. He stops playing for a moment and uses his smartphone to search online for a place to dine. Andrew is creating GPS data that records his current location[2] and the fact that he is looking for a restaurant. The app provider, to generate advertising revenue, makes this data available to third parties, who in turn use it to place their marketing offers and events. So, Casino Z could be watching for these searches and sending locationally targeted marketing events to Andrew. Figure 1 shows a screen shot of the Foursquare application Andrew used for his search. As you can see, he is shown several location-based offers in his vicinity—most of which are not at Casino Y.[3]

Figure 1: Location-Based Offers on Foursquare

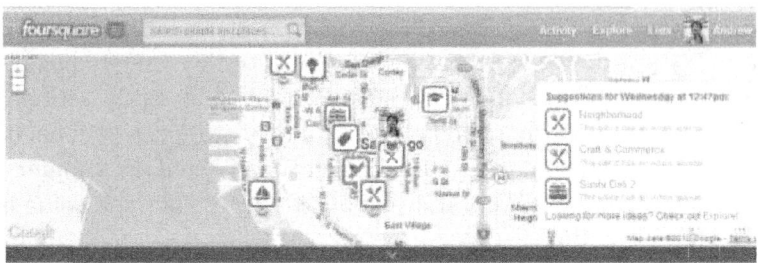

This interaction data4 is even more interesting because it is managed by another party, Foursquare, which will provide advertising services to almost anybody who is willing to pay. This interaction data is therefore very different than the transaction data that drove our marketing efforts in the last century: Our competition has access to the same resource.

While Figure 1 shows the customer placed on a street map, which is the traditional spatial view, with the introduction of indoor-capable locators, we can now see exactly where we are inside a retail space or casino. One does not need to look much further than Google Maps to see some of the latest work in this indoor space. (See Figure 2)

Figure 2: Inside Spatial View from Google Maps

Figure 2 shows how merchandise is laid out inside a retail store and tracks the customer's movement as they navigate the retail space. Once again, this data does not belong to the owner of the retail space, but to the individual customer—and is often shared with the app provider, in this case, Google Maps.

Where Are My Customers?

It is well established that proximity to competition is one of the primary drivers of business, and now census data can give us reasonably accurate market share numbers. With the 2010 census data now available, we can see the number of potential gamers by census tract across a market catchment. The knowledge of where the customers are is interesting, but what is even more fascinating is the relationship that this has to the catchment of competition. Quite simply, we can now see the effect of competition on our market share in terms of both visits and revenue. Taking this one step further, we can see big data effects in many cases. For example, if the customer uses our Casino X application to search for upcoming shows or events, we can see where that customer was when they did the search. We can now see the relationship between where our customers searched and interacted with our property, and where they were located.

Location stands at the center of this analysis, with its special ability to link seemingly unrelated data sets, such as census information and online searches. To manage this data and these relationships, we need locational storage, locational query, and locational analysis tools. These tools form the backbone of the locational analysis systems of the future.

Collecting the Data

To begin to make sense of this new frontier in gaming data, we need to understand what data we can, and cannot, collect about our customers. With the advent of locational services on mobile phones, we can ask our customers to opt-in to sending us locational data, accomplished via a casino app. So, assuming we are able to design an application that engages a significant portion of our customers, we can begin collecting this important new data stream. Many casinos are already doing this, presenting apps that not only describe the various services the casino has to offer, but also interacting with the customer and making them eligible for additional offers when they use the app. Basically, casinos are now doing via smartphone what they've done for decades—bribing customers to encourage increased engagement.

Now we have an increased wealth of data about our customers. In addition to all the transaction data we get while they are on our casino floor, we have locational data whenever they use our mobile app. This explosion increases the dimensionality of our data exponentially, as discussed in our last articlepart.[5]

Mining the Data

Enter the world of predictive modeling. With all this gaming data and locational data, we need to figure out how to extract value from it. Let's think about what we have to work with, and what we are trying to accomplish.

We know, or should be able to know with some effort, the following types of information:

- Gaming behavior – We can track every transaction on our gaming floor.
- Retail, F&B, and entertainment behavior – We can track purchases in our non-gaming outlets.
- Basic demographic information – Age, gender, address and other contact information.
- Appended demographic information – We can purchase additional information about our guests, such as income, magazine preferences, presence of children in household, car purchases, etc.
- Locational data – We know where they are when they use our mobile app.

With this information, we can start to ask questions. Where are our customers when they use our app, and what are the consequences of this action? Do they use the app when they are near the competition, then come to gamble with us? Do they use the app when they are on property? Does using the app increase or decrease their play? And how does all of the other data (non-gaming, demographic, etc.) change how the customer's interaction with the mobile app relates to their gaming behavior?

These questions are all answerable via predictive modeling. To build a predictive model, we need input data, output data, and a lot of outcomes. Let's tackle one of our questions as an example: Does using the app increase or decrease customer play? In a simplified version of this model, we first need to organize the data in a way that the above question can be answered by a computer model.

First we input our data. We have loads of gaming data, but we are trying to determine if the use of the app increases or decreases customer play. So, for each customer we need to collect all the gaming data prior to his or her use of the app. In addition to the gaming data, we have other behavioral data such as retail purchases, and again, we need to collect this information prior to the use of the app. On top of all this, we have basic and appended demographic data, which is usually time independent. And, finally, we have the location(s) where the customer used the app.

From this soup of input data, we also need to determine our output data. For this simplified example, let's attach a "1" to any player whose average daily play increased after using the app and a "0" to any player whose average daily play did not increase. This gives us what is commonly called a base modeling table. From this table, people who are experts in predictive models can weed out which metrics are predictive of a customer increasing his or her play after using the app, which metrics are irrelevant, and which weights to apply to the relevant metrics.

Now the fun starts! It's time to apply the model. When doing this, we need to keep in mind that when a casino rolls out an app, it often takes time for customers to adopt usage of the app. In the beginning, a small fraction of customers, perhaps 1 or 2 percent, will actually use the app. Even as adoption grows, this percentage will likely remain below 50 percent for years to come. Nonetheless, for our current customers, we can learn which ones are worth encouraging (strongly, via offers) increased app usage. We can also determine when to use the app to push offers based on whether the customer is close to a competitor, close to our casino, or whatever our models tell us is relevant.

In addition, we have non-customers that we want to take from our competitors. The predictive modeling described above can assist with this, too. For non-customers, we can tweak our models to remove gaming and on-property non-gaming data, and then understand how demographic and locational data can combine to tell us how to best leverage mobile tools such as Foursquare—

getting the right offer to the right person at the right time. These models can be taken further by combining broad sweeps of Internet-trawled data with companies such as PredictivEdge6 to build predictive analysis on this data.

Finding the Money

People need to eat, people need to socialize, and people need people. We need to look no further than the movies to see how people still choose to go to the cinema and enjoy a night out rather than stay home. As we have illustrated, location can tie our customer interactions together with the broad world of social data. Clearly, location is a truly differentiating feature of a brick-and-mortar business, and it is time to move your brick-and-mortar data into the GPS world.

1 "Gaming Interactions: The Invisible Force of Social Networks," Singh and Cardno, CEM, February 2010. 2 Inside data is less accurate, as it is often based on WIFI locational services.
3 Screen shot taken from Foursquare, February 2012.
4 "Gaming Interactions: The Invisible Force of Social Networks," Singh and Cardno, CEM, February 2010. 5 "Where's the Money? Part 9," Cardno and Thomas, CEM, March 2012.
6 Reference used with permission of Bill Thompson. www.predictivedge

CHAPTER 11: WAR ROOM ANALYTICS

Authors' Note: This is part 11 of our 18-part series, "Where's the Money," and part three in our subseries on big data. In this article part, we look at how money is made by enabling operators to act and that these actions can be facilitated by war room analytics. We then examine how reporting and analytics are very different, and how confusing them is like confusing a motor bike with an 18-wheeler; both can get you to your destination, but the ride is quite different. We describe the art of analytics and how it draws on creative insight and skills that enable a world of collaborative opportunistic profit improvement. Looking deeper into this, we draw on some of the previous case studies of gaming analytics to illustrate how this collaborative opportunistic has been successfully applied.

War rooms are places where teams meet to collaborate. They have existed for centuries and have been, as the name suggests, central to the intelligence activities of many wartime campaigns. The walls are typically plastered with initiatives that are central to the theme of the war room, as war rooms are the end point for what is often a huge data collection process, and information consumption is almost always a team effort. This article part examines how war rooms can and should play a critical role in the intelligence of business operations and how they are very different from other analytical approaches.

With the explosion of smartphones and tablets over the past few years, software companies have been racing to find ways to present analytics in smaller and smaller packages. A search of any app store will uncover dozens of analytics applications for mobile devices. Judging by all the hype, it appears that this is the wave of the future—a world in which people have their heads buried in their smartphone all day long, trying to uncover hidden truths about their data and emerging at the end of the day with brilliant insights to help drive their businesses.

There is only one problem with this idea: analytics and reporting are not the same thing. But rather than launch into a Webster's Dictionary-based discussion of the differences between the two, let's consider slot optimization as an illustrative example. (See Figure 1) As we described in "Where's the Money? Part 6" (CEM, December 2011), the goal of slot optimization is to increase the player experience in ways that drive incremental gaming revenue. There are many differences between optimization metrics such as utilization and spend per hour and outcome metrics such as win per unit and hold percentage. The differences between these metrics lie parallel to the differences between reporting and analytics.

Figure 1: Analytics vs. Reporting

Analytics	Reporting
Step 1: We want to know what changes we can make to our slot floor to achieve our goal of increasing the player experience in ways that drive incremental gaming revenue. To accomplish this, we first need to know what the optimization metrics (e.g., utilization, spend per hour and devotion) are for every game on our floor. In other words, the desire of the analyst is to increase revenue through optimization.	Step 2: After doing slot optimization, we need to know if we were successful. In particular, were we able to affect win per unit (an outcome metric) in ways that were incremental? At a high level, one can measure the results by looking at, for example, year-over-year gains in win per unit. In other words, the desire of the consumer of the report is to understand the outcome.
Step 3: The report in Step 2 may be skewed by other external factors, such as a recession or a booming economy. In order to see past this, one needs to delve deeper into the data and understand the impact made by the changes to the slot floor. In particular, one needs to look	Step 4: After digging deeper into the analytics and providing a better measure of incrementality using the analysis described in Step 3, one can summarize this data and demonstrate the effectiveness of the slot optimization. Again, these

not only at the games changed directly, but also at the cannibalization effect on other nearby or similar games.	reports are reporting on the overall results and outcome.

As we can see in Figure 1, we have two examples of analytics and two examples of reporting. These examples show how analytics is focused on optimization and how reporting is focused on the outcomes or results. There are a few other factors regarding analytics vs. reporting that are important to be aware of; namely, that analytics is big, analytics is complicated, analytics is not an exact science, and analytics is a team effort.

Analytics is BIG

Step 1 (from Figure 1) entails pouring over thousands of data elements, trying to find areas of opportunity to improve the customer experience.

Analytics is Complicated

Step 3 requires measurement of the impact of perhaps hundreds of games and attempts to remove the biases created by seasonality, changes in customer preferences over time, and cannibalization. Reporting, on the other hand, is much simpler and smaller, and thus lends itself well to miniaturization to a smartphone or tablet. We propose that analytics should not be miniaturized, and will make the case below through our argument for the creation of an analytics war room.

Analytics is an Art

You heard it here first! Analytics is an art, not a science. Reporting is a science. If you want to know how many widgets you sold yesterday, you take your source system tracking data, ETL move it into a data warehouse enterprise data asset, push the data into a front-end business intelligence system, then access that system via a computer, tablet, or smartphone, and—voila!—you know how many widgets you sold yesterday. However, if you want to know how you can drive incremental widget sales, the task becomes much more difficult.

Let's return to our example of slot optimization. Say we want to introduce a hot, new game to our floor. We have to first determine which games to remove in order to make room for this game. In the old days, slot operators would simply cull the games that had the lowest WPU. However, we've since learned that many other factors are important:

1. Should the new game be placed in a high- or low-traffic area?
2. Will the new game cannibalize nearby games?
3. Will removing old games cause us to lose customers who are loyal to those games and are now frustrated by their disappearance?

There are metrics that can help us answer all of these questions. However, there is no set formula that will tell us the exact location where the hot, new game will drive the most incremental gaming revenue. Rather, we have to use our analysis of past moves to understand how customers react to changes we make on the slot floor—and this is a complicated exercise involving multiple metrics on dozens of changes made to thousands of games. In this way, analytics is a healthy combination of both art and science.

Analytics is a Team Effort

It our belief and experience that action is what drives value from analytics, and our experience also shows us that action normally requires a team effort. Collaboration regarding analytics is about meeting as a team, deciding which actions to take, and then taking those actions. This collaboration then extends to the operationalization of decisions made.

When looking at optimization metrics, the analytical approach is innovative and insightful; when looking at reporting outcomes, we need a structured review. Figure 2 summarizes the relationship between types of metrics and the collaboration required to act on those metrics.

Figure 2: How Optimization Metrics Relate to the War Room

	Metrics	Collaboration
Reporting	Outcome focused	Individual centric, business monitoring focused
Analytics	Optimization focused	War room centric, opportunistic focused

The War Room

The next question is: "what is the best setting for the analytics team?" Enter the war room.

But what should the war room look like? Should it be virtual, with team members sitting behind their own computers on a conference call? Should they all physically be in a room together, but tapping furiously on their own tablets or laptops? Or looking at a PowerPoint presentation displayed on a projector? We personally agree that none of these options are conducive to team analytics.

Instead, we believe and have discovered in practice, that the best approach is to go big. One great way to achieve this is to make massive printouts using a plotter, which allow you to step back and see the big picture, then zoom in (by shuffling your feet forward) to see the details. Hang these printouts on the walls of your war room and multiple people can look, point, write, and debate, making the best use of the information available.

Going back to our slot floor example, a great printout to make would be a heat map of your slot floor, with all of the necessary metrics written in detail as well. Each slot machine gets represented by a square on your map. Each square can then be colored based on whatever metrics you choose—and it is possible to use more than one color at a time. For example, you could have a color that represents utilization and a color that represents win per unit. Inside each square, you can then include more detailed data, such as the name of the game, the game type and cabinet, the utilization, win per unit, spend per hour, devotion score, and hold percentage. With this approach, you can stand back and look for patterns in the heat map, then move closer to see the detail in each game. This is one of the strategies put into action at the real-life analytics war room at Silverton Casino.

Troy Freet, Teradata database administrator at Silverton Casino, says, "The war room has become a central focus for our analytics. Combining our atomic level data with high-resolution visualizations allows the users endless flexibility and interaction in a highly collaborative environment. The analytic environment provided by the war room has drawn senior management into the analytics process with very positive results."

The best war rooms will make the analytics immersive (in that you physically walk into the room) as well as collaborative by holding strategy meetings in the room. Drive actions from your war room by making sure that decision-makers are being fed both the collaborative and the creative insight opportunities to make good decisions.

Show Me the Money

According to Thomas H. Davenport: "Analytics competitors are more than simple number-crunching factories. Certainly, they apply technology—with a mixture of brute force and finesse—to multiple business problems. But they also direct their energies toward finding the right focus, building the right culture, and hiring the right people to make optimal use of the data they constantly churn. In the end, people and strategy, as much as information technology give organizations strength."1

As Davenport so clearly says, people and strategy are central to organizational strength. And while there are likely to be many paths to this strength, war rooms focused on optimization metrics are a powerful way to bring forth strategies and to unite people in a culture that enables change and innovation. This is particularly relevant when we consider the turbulence in business today, whereby innovation is ever more important.2 In weathering this kind of turbulence, analytics has become a central part of many decision-making processes and businesses. In addition, operators are often faced with the difficult task of throwing away hard-earned business practices to innovate in different directions entirely. The role of analytics is central to the discovery of these new directions and their team-based implementation.

Teradata Database Administrator Troy Freet in the analytics war room at Silverton Casino.

---

1 "Competing on Analytics," by Thomas H. Davenport. The Harvard Business Review. Retrieved from www2.mccombs.utexas.edu.
2 "Innovation in Turbulent Times," by Darrell K. Rigby, Kara Gruver and James Allen. The Harvard Business Review. Retrieved from http://hbr.org.

CHAPTER 12: MAGNET GAMES AND PARADISE FISHING

Authors' Note: This is part 12 of our 18-part "Where's the Money?" series. In this article part, magnet games will be examined and their role in mini casino design explored. Magnet games often dominate the character of both the area and the players in an area. The examination will continue with a deep dive into the performance characteristics of Aruze's Paradise Fishing, using preference filters and market basket analysis.

Magnet games are leaders in slot performance. They are highly correlated to their surrounding games, and they also outperform the surrounding product. These games are wonderful additions to a gaming floor, as they both attract players to a game, and they might also create "spill," driving additional play on surrounding games. If a game has a high spill effect, then we have truly found a driver that can be used to optimize the gaming floor. Furthermore, we can measure this spill effect as a "lift" on surrounding games by looking at reverse cannibalization. To do this, we need to understand these two new game metrics of spill and lift, which we can measure by performing a directional analysis.

Spill is the amount of additional revenue that a game contributes to an area due to overflow from its high demand. As additional games are added, the spill will typically reduce. Spill can be measured in dollars of theoretical win. Lift is the percentage of increase in revenue in the surrounding games.

Directional analysis is a probabilistic view of the migration of gaming floor play that shows the likelihood of a player moving from the magnet game to another game. Think of this as a kind of chaining process where the customer's current play pattern has an increased probability of driving the player to a surrounding game. This directional analysis provides a very insightful view of the gaming preference. Quite simply, the questions are, "From this game, which game are customers likely to move to next?" and "Where did customers who are playing this game likely play previously?"

To answer these questions, we need to look at player preference. There are two ways of doing this: The first is game preference, which is behavioral and observable, and the second is an actual loyalty for the game. During each visit that a new player exhibits a preference for, a game gives us clues that the player has game loyalty. (See Chart 1)

Chart 1: Player Preference and Loyalty

(A) Game Loyalty	Game loyalty is a true dedication to the game. The key characteristic of game loyalty is that the player will seek out the game if it is moved. Games with high game loyalty can be moved, and the loyal players will move with the game. Preference play for a game is a clue that the player has developed game loyalty.
(B) Preference Play	Preference play is calculated from the percentage of play on a specific game or game theme during a player's visit. It does not mean that the player is loyal to the game—it just means that the player spent most of his or her time on the game.

Understanding if a player has game loyalty provides predictive knowledge about many aspects of game optimization. If we are intending to move a game, players who have game loyalty need special consideration, and possibly specific communication, about the movement of the game. Also, games with loyalty may present an opportunity to become the centerpiece of a mini casino strategy.[1]

One of the major differences between loyalty to a game and preference for a game is the amount of time it takes to measure the two. Preference for a game can be measured on a single visit—with the understanding that this preference can change over time. Loyalty to a game cannot be measured on a single visit, or even a handful of visits. However, loyalty is no doubt measurable. For example, we can look at the effects of physical movement of the game and measure the number of players who are prepared to seek out the game in its new location. This brings us to the very interesting challenge of calculating the probability that a player is truly loyal to a game if they exhibit game preference.

Bayes Theorem

We will now show how game preference can be used to calculate the likelihood of game loyalty using the Bayes Theorem. The Bayes Theorem is a wonderful probabilistic model that gains knowledge from clues or hints. Let's calculate the probability of game loyalty (A) given the evidence of preference play (B) on a player's visit. The probability of A given B is equal to the probability of B given A times the probability of A divided by the probability of B:

$P(A|B) = P(B|A) * P(A)/P(B)$. (See Chart 2)

Typically, the event that we are interested in is not observable, but the evidence of these events is observable.

Chart 2

Notation	Information	Observable	Illustrative Probability
A|B	Probability that a player has game loyalty given they show preference play on one visit.	No	=60%*10%/20% = 30%
B|A	Probability that the player gives preference play given they are known to be loyal to the game.	Yes	60%

B	Probability of prefer- ence play on the game on any visit.	Yes	20%
A	Probability that any player is loyal to the game.	Yes	10%

In other words, if a new player shows preference play for a game, in our illustrative example, they are more likely to have game loyalty. Understanding that this new player has game loyalty means we can say that there is a 60 percent chance they will play on the game during future visits. Further visits provide more clues about the player's feelings toward the game, increasing the accuracy of the inference model.

Paradise Fishing

In our series of article parts about Jackpot Wharf2, we described how this mini casino was recently completely re-merchandised and reconfigured to draw in more visitors from the adjacent Bass Pro Shops at Silverton Casino. The Jackpot Wharf initiative continues to maintain its revenue increase at around 50 percent above pre-change numbers. Paradise Fishing is a popular penny video slot game that features an impressive large screen community bonus. Paradise Fishing has extremely high occupancy and is a strong financial performer. As such, it is even more important to merchandise the immediate area correctly to take advantage of the spill opportunity it provides.

The following section analyzes the preference play on Paradise Fishing to enable changes that will improve the spill effect onto surrounding games. To do this, we need to start with a preference filter.

A preference filter is used to screen out the noise from your data. It makes sure that only the gaming activity that your customers spend on games they like shows up in your analysis. It ignores the trial transactions—customers trying a game and not liking it, customers spending a few minutes waiting for the actual magnet game to open up, etc. A game can show good performance because of heavy trial activity, but this can quickly fall off when it fails to retain customers. A preference filter is one simple way of developing a model for preference play, and it does not filter out players; it filters out some of the play of all players, in essence leaving only preference play.

The inspatial graphic in Figure 1 shows the gaming activity for Jackpot Wharf players who showed moderate interest in Paradise Fishing. We wanted to look at this group of players to test our merchandising strategy—we were looking for fairly even play patterns inside Jackpot Wharf. Without preference filters, it appears we have succeeded at our goal: Paradise Fishing players are playing throughout the entire area, and we might assume that it is properly merchandised.

Figure 1: Paradise Fishing without Preference Filter

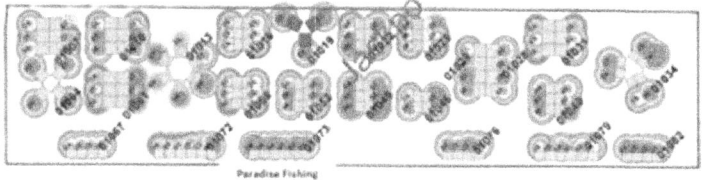

But see what happens when we apply a preference filter in Figure 2.

Figure 2: Paradise Fishing with Preference Filter

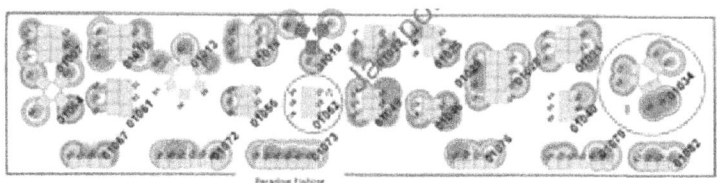

Suddenly, entire banks are gone! Specifically, see Bank A, which in the original graphic, looked very popular. After applying the preference filter, not a single Paradise Fishing player shows preference play on these games—and that's our target group for this merchandising strategy! However, as Figure 1 shows, many of them did play on the bank. Possibly the players were merely biding time while waiting for Paradise Fishing to open up. This is a potential high performing location because of its proximity to the Paradise Fishing magnet bank.

Also look at Bank B. In Figure 1, without the preference filter, the bottom two games were the weakest on the bank. But once we apply the preference filter, we see that this group of customers clearly prefers the bottom games. As with Bank A, many customers played the top games, but the majority of the group showed preference play on the bottom. Preference is an extremely important distinction, as product in a high-traffic area may perform well, especially if it is transient traffic.

We can now apply the same preference filter to the entire floor to locate products that this target group prefers to play, and then bring it into the area. Further, we can isolate players who actually prefer Bank A and identify an appropriate area to move it to. (See Figure 3)

Applying the preference filter floor wide also shows a handful of other preferred products that the target group plays heavily. The majority of their other preferred products are actually on the opposite side of the casino floor. It would make sense to pick up one of these banks and simply move it right next to the magnet games.

Applying preference filters, we can find a new home for Bank A. Figures 4 and 5 show the preference filtered market basket numbers.

In Figures 4 and 5, we are showing "tree map" data visualization. This visualization is very good at handling multiple hierarchical dimensions of data. In this case, each small box represents one slot machine and the size and the color both represent the preference filtered revenue numbers. These small boxes are grouped into larger boxes based on their attributes, for example in Figure 4, we see that Video Slot is a large box that encompasses a large number of small boxes (or slot machines), and that within the Video Slot area the House section is the largest, meaning that house games contain the majority of the preference filtered revenue. Simply put, the tree map is a visual market basket analysis.

Figure 3: Top Preferred Games

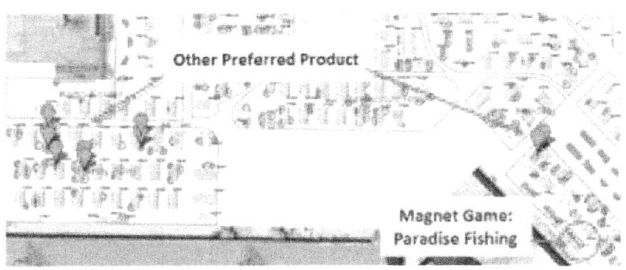

Using preference filters and tree maps, we can see that customers who are moderately interested in Paradise Fishing have much different product preferences when compared to customers who are moderately interested in Bank A. From this, we immediately see that Bank A players are a lot less inclined to play video slot participation products. They are most likely to play house WMS video slot games, some of the older IGT penny titles and Bally's Blazing 7s.

Just as we did with the target group, we select Bank A's most preferred games and are able to locate an area with decent (if scattered) activity and several non-preferred banks. This area should prove successful for Bank A. (See Figure 6)

Figure 4: Game Preference Paradise Fishing

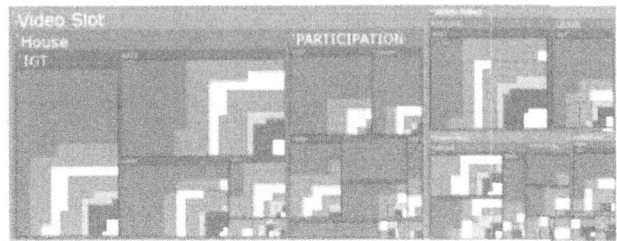

Figure 5: Game Preference Bank A

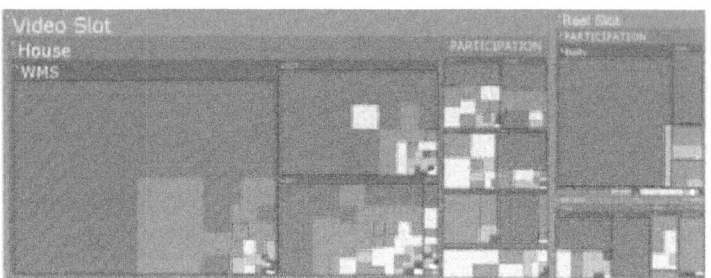

Figure 6: New Location for Bank A

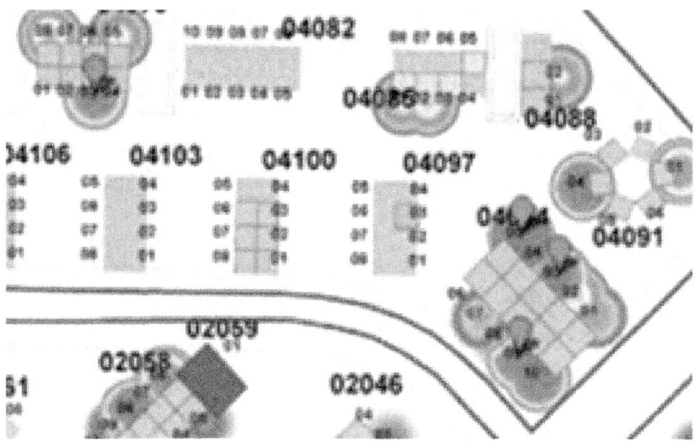

Where is the Money?

The numbers on Jackpot Wharf are hard to argue with: An ongoing 50 percent increase in net revenue resulted from a focused and well-informed strategy of gaming floor optimization.3 As we have illustrated here, Jackpot Wharf has even more opportunity for improvement, and the decisions about these games are simply not based on outcome metrics—they are based on optimization metrics. We showed how preference filters are a powerful way of conducting gaming floor optimization and that traditional outcome metrics would have missed these opportunities altogether.

1. CEM January 2012, Cardno, Thomas, Evans: Jackpot Wharf, Part 1
CEM February 2012, Cardno, Thomas, Evans: Jackpot Wharf, Part 2
CEM December 2010, Cardno, Singh, Thomas, Evans: Penny Alley, Part 1
CEM January 2011, Cardno, Singh, Thomas, Evans: Penny Alley, Part 2
CEM February 2011, Cardno, Thomas, Evans: Penny Alley, Part 3
CEM March 2011, Cardno, Thomas, Evans: Penny Alley, Part 4
2. CEM January and February 2012, Cardno, Thomas, Evans: Jackpot Wharf, Parts 1 and 2
3. CEM December 2011, Cardno, Thomas, Where's the Money, Part 6: Player Experience and Slot Optimization

CHAPTER 13: GREAT GAMES IN GAMING— WHEEL OF FORTUNE

Authors' Note: For the lucky 13th installment of this Where's the Money series, let's dig into the performance of another great game in gaming—in this case, the classic IGT game Wheel of Fortune™ (WOF). This amazing product is considered a must-have in nearly every casino across the planet, so we'll dig into how the analytics related to this "must-have" perception, in fact, make it a reality for at least some properties. Then we'll expand on strategies for capitalizing on WOF to drive revenue by building out its ability to be a central feature or magnet game for an area.

Wheel of Fortune is one of the great games of gaming, and it is sometimes called "the most popular games of all times."[1] This remarkable game has continued to evolve over its 16-year life span and has continued to command revenue share. Two examples to illustrate its continued popularity are the anniversary release in 2006 of Wheel of Fortune™ Special Edition[2] and winning the first place award in 2011 for "Favorite Progressive Slot" in Casino Player magazine's Best of Gaming Awards.[3]

Now, with player favorites like Wheel of Fortune in mind, let's explore return on investment models that categorize games based on their long-term value—and the long-term value to the gaming operation. Wheel of Fortune is an important game, and we will show how its revenue model fits into a category with other long-lasting themes on the gaming floor. We'll also show how the continued popularity of this gaming product is a reflection of its ability to continually attract new preference players.

Revenue Growth vs. Revenue Protection

We have talked extensively about revenue growth through strategic, data-driven changes to the slot floor. While we believe many of the methods discussed are new and innovative, the concept of growing revenue by changing a slot floor is decades old. However, as the subtitle of this section suggests, growing revenues is not the only job of a slot operator. There is a second task that slot operators should always be aware of, especially in light of today's competitive environment: revenue protection.

When tasked with revenue protection, the goal is to protect, or retain, patron share of wallet. There are many decisions that are made in the name of growing revenues from one patron segment that inadvertently reduce revenues from another segment. In the aggregate, this may be a positive outcome, a wash, or even a negative outcome. And when digging deep into the results of your slot changes, the analyses may show that outcomes that look positive at first might actually cost you revenue.

As a simplified example, let's suppose that we have two games. As shown in Figure 1, Game A is performing at $200 win per day and Game B is performing at $100 win per day. Because of this, the slot operator decides to remove Game B and replace it with a second machine of Game A, as shown in Figure 2. The second Game A does $150 win per day, and while the operator wonders why it didn't also do $200, he's happy enough with the extra $50 per day. But what if the original Game A is now only doing $150 as well? This deeper look shows that the change was a wash, but the analyst may conclude that "I'm still getting $300 per day out of the two games combined, so at least I didn't hurt anything."
However, this analysis of the game change can change dramatically when taken from the patron's perspective. Expanding the example to consider patron preference, we realize that patrons who prefer Game A are pleased with the change, as they now have two machines to play instead of only one. However, we haven't truly changed their gaming experience—they are still playing Game A. In addition, outside of busy times when the first machine is occupied, there is no real benefit to these patrons.

Now let's consider patrons who prefer Game B, whose preferred game is completely gone. Game B patrons have had a massive impact on their experience—they can no longer play their game of choice at the casino. Their business may be lost both now and in the future.

Fast forward to 3 months after the change and consider what happens to our illustrative example next. Patrons who prefer Game A may start getting bored of the game, reducing the demand back to a total of $200 per day, which is now being split between the two machines. Patrons of Game B are gone. If we hadn't made the change, we'd still be getting $200 on the original Game A, and we'd still have play on Game B. In effect, by replacing Game B with Game A, we lost 33 percent of our future revenues.

Figure 1: Game A and Game B Before Change

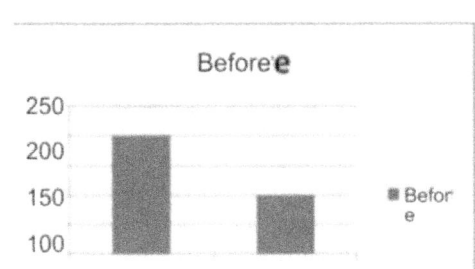

Figure 2: Game A (x) After Change

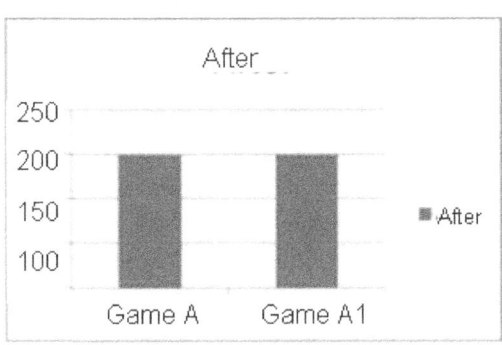

Clearly, slot operators must be equally mindful of two often competing goals: revenue growth and revenue protection.

Adding the patron wallet dimension can, and in our experience often does, change the analysis dramatically. Quite simply, our goal is to first grow slot revenues through new business, and second, to protect revenues by keeping patrons satisfied, thereby preventing them from defecting to other casinos. As both our experience and our illustrative example show, short-term yield improvements may result in long-term revenue costs.

Decay Curves and Patron Growth

Individual game themes have their win per day decay over time, and different games can have quite different performance decay curves. This performance decay happens at quite different rates for different games. Consider, for example, the Wheel of Fortune games. These games have been around since 1996 and are still in relatively high demand. In fact, it seems like demand for this product is never-ending.

Contrast this to many other titles (possibly those that are gimmicky or based on movies), where after a year or two, the play drops precipitously. In this manner, we can look at our floor and forecast what our performance would be if we never changed out our product. We can imagine a small group of games that maintain their play and a much larger group of games that decline, often quite quickly, over time. Table 1 shows a basic classification model for game performance that describes different kinds of performance decay curves.

Table 1: Categories of Performance Decay Curves

Decay Curve	Description
Never ending	These themes are continuing to attract new players, and manufacturers often continue to release new versions or upgrades. One example is Wheel of Fortune.

Long-term following	This group of themes has patrons who are super loyal and who will stay with a game for many years. However, these themes are not picking up many new players, so their performance is gradually decreasing. One example in this category is some versions of keno.
Box office hit	These themes are hugely popular when they first appear on the gaming floor, but after a short period, their performance hits a wall. Some movie-based video slot products exhibit this curve.
Never made it	These themes never seem to make it. They start low and stay low.

Looking at the decay curves in Table 1, one of the key differentiators between the categories is the ability of a gaming device to attract new players and expand its popularity. Players in many ways flow into and out of the business, and it is the job of the gaming product to redirect this flow to maintain or grow revenues. Keeping the floor fresh or maintaining a constant supply of more attractive products results in a growth in the gaming revenue by maintaining or growing the patron base. This maintenance of patrons is a very complex issue, but our intuition indicates that fresh and high-preference gaming product, despite often higher performance numbers, are not driving additional short-term revenue, but are in fact driving medium-term patron growth.

We have created what seems to be a contradiction— according to the decay curves, we should constantly refresh our gaming floor, but the illustrative example indicated we should keep older games even if they are underperforming. Which approach is correct?

In fact, both recommendations are sound. The contradiction exists because we have so many types of patrons, each with their own gaming experience preferences. Consider the patrons who love Wheel of Fortune, for example. We can study the other games that these players prefer to play. It is our experience that players who play Wheel of Fortune do play other games, typically other progressive games. The big question is whether it is Wheel of Fortune that "makes" this category of player by continuing to capture the attention of new players into this type of gaming experience.

To round out our discussion, let's consider two different patron types: "traditionalists" and "change seekers." Traditionalists are patrons who love the games that first gave them that rush of gaming or that remind them of a big win they had months or years ago—or perhaps they are just stuck in their ways. Change seekers, conversely, are always looking for the newest, best, novel experience. They want something different and exciting when they choose which slot to play.

These change seekers require a completely different treatment than our traditionalists, so when looking at these types of patron groups, it becomes the job of the slot operator to understand how to protect and grow gaming revenue for different segments of patrons. This could involve retaining games that are loved by the traditionalists, while at the same time bringing in new and exciting product that motivates the change seekers to keep coming back to the casino.

Where is the Money?

Now that we have explored gaming machines and their relationship to patrons, we know that the gaming machines with very high performance numbers might not be increasing overall revenue; this is especially the case where the market is oversupplied with gaming product. Instead, this gaming product is producing medium term patron database growth measured in the ability of the property to grow different segments of gaming customers based on their product preference.

There is little doubt that getting the right mix of gaming machines continues to be a central theme of finding the money on your casino floor. This chapter looked at what has often been judged as one of the most successful slot games, Wheel of Fortune, and explored some new ways of looking at the value of this and similar games that extend beyond the traditional numbers (such as compared to house average). This discussion is the beginning of a dedicated math model that will show the return on investment of a gaming machine that is based, in part at least, on the preference for that game and its ability to attract new patrons over the medium term.

1 Extracted June 2012 from http://en.wikipedia.org/wiki/International_Game_Technology#cite_note-8.
2 Refer to the press release at www.igt.com/company-information/news-room/news-releases?NewsID=753618.
3 Refer to the press release at www.igt.com/company-information/news-room/awards-and-honors.aspx.l
4 Extracted June 2012 from www.pewsocialtrends.org/2008/12/17/who-moves-who-stays-put-wheres- home/.

CHAPTER 14: GREAT GAMES IN GAMING—CLUE™

In this installment of our "Where's the Money?" series, we dig into the analytical challenges posed by another great game in gaming, WMS' Clue™. This great game combines an online experience with a traditional in-casino experience. The game is changing the way that players play, and the analytical challenge is now how an operator can understand player behavior when much of that player behavior is interaction with a "gaming" experience outside the four walls of the casino. But before we dig into that, let's first take a look at the data we have to work with.

In the past, the gaming industry has had near complete, albeit inaccurate, data. With this data, operators have trawled for customer preferences, piled those preferences to build market baskets, and studied patron frequency patterns. This detailed data about our patrons has been at the core of many activities in gaming for at least the decade. Across industries, this data is of significant size and is, in many ways, comparable to the retail industry.

It may seem easy to underestimate the volume of data in the gaming industry when compared to other industries such as the retail industry. It can be tempting to consider the data in the retail industry as being significantly larger. However, the following analysis shows that customer-related data is much more granular in the gaming industry. Of course, the real difference is that retailers have many more stores than casino operators have casinos—there are 4,500 Wal-Mart stores.1 So as gaming operators, we have similar granularity of data, although oftentimes we have a smaller size of data. (See Table 1)

Table 1: Gaming Data vs. Retail Data

	Comparison	Retail	Gaming
Space	Similar	Retail stores range in "typical" size from a 100,000-square-foot grocery store to a 350,000-square-foot big box format.2	Gaming floors are typically approximately 100,000 to 200,000 square feet (with 1,000–2,000 slot machines).
Products per square foot	Similar	A single bay (or part of a shelf), might hold one to 50 products.	A single slot machine can take many configurations, but five to 50 user options are common.
Transaction data	Retail is greater	The transaction data is granular, with each product type being sold defining the base level of the data.	The data is typically monitored every 15–60 minutes and is
Customer data	Gaming is greater	Customer information is typically one transaction per visit. Customers are often "unknown."	Patrons generate many time and space tracked events during each visit, with a minimum of one transaction per card per visit.

Gaming Becomes Social

But how does this connect to Clue? Let's take a look. The July 2, 2012 edition of the Las Vegas Sun summed up the latest from the social gaming boom: "Caesars Entertainment bought Playtika, the Israeli developer of Slotomania, a Facebook slot machine game. MGM partnered with Playstudios to launch my Vegas, a Facebook application coming this summer that allows people to play blackjack or slots. Boyd Gaming and MGM invested in bwin. party." So gaming operators are clearly actively perusing online gaming. And in actuality, it seems like many gaming operators are not only entering online gaming, but are oftentimes buying companies that "manufacture" these online gaming experiences.

In addition, slot machine "manufacturers" are actively entering the online gaming world. The IGT interactive website states how the company has "10+ years in ... online and mobile." This, combined with Nevada's approval of online poker, moves IGT firmly into the world of being an online gaming operator.

As was mentioned in "The Demise of the Slot Manufacturer", the gaming industry can be compared to the software industry in that both have huge growth potential and huge change potential (see the July 2008 issue of CEM, available at www.aceme.org). In the software industry, the growth came from unexpected places—the PC and now the mobile platform. In gaming, one approach is to combine the online and in-casino gaming experience, which finally brings us to our great game of Clue™. The WMS Clue game brings benefits of the online gaming world to the casino.

A Great Game

Clue is known to be a strong-performing game, but what is exciting about Clue is that the player can leave the casino, drive home and then play online at www.playerslife.com to accumulate features that will change the in-casino gaming experience. While we cannot predict if this model of game play will be a long-term winner, it definitely creates a new kind of analytics challenge and a new kind of data.

Key Attributes

We have "field tested" Clue in order to learn its key features. Following is a list of observations based on our casino play and subsequent online experience:

1. The game, like many other premium games, relies heavily on its bonus features.
2. At the casino, initially certain "rooms" in the Clue house are locked. If a customer receives a bonus round, it might take place in the kitchen or in the study, while the billiard room remains locked.
3. Customers can create a login ID and password at the device so their progress can be tracked on the Player's Life website.
4. The slot machine prints a ticket, directing the player to www.playerslife.com.
5. Customers can unlock rooms during their play at the casino.
6. Customers can go home and unlock additional rooms via an online experience.
7. Online customers have a wide array of options, including tracking the "trophies" earned for completing various achievements (e.g., unlocking a room in the casino); playing "Quick Clue" to temporarily unlock rooms in the casino game (see Image 1); and visiting a forum to discuss their experience with WMS games. These forums are quite robust—the forum for Clue was 45 pages long, with comments such as "Great Game!", "Will play again on my next visit, "Birthday party in Vegas in 5 days!" and "Is this game available in Oklahoma?" There is also a leaderboard to track who has won the most credits in the casino, listing the player and the casino.

Image 1: A Player's Life Screen

The Invisible Force of Social Networks

According to the July 2, 2012, edition of the Las Vegas Sun, "Close to 1 million people have signed up for Player's Life. A million more have registered over the past six months for WMS' Facebook slot game 'Lucky Cruise.' Last week, WMS acquired Phantom EFX, which develops interactive and mobile slot games."

Without social media data, social media interactions are an invisible force, an idea explored in "Why Do I Need Math? Part III, Gaming Interactions: The Invisible Force of Social Networks" (see the February 2012 issue of CEM). According to the articlepart, this force could be so great that players can be researching and playing online, generating "a huge wealth of behavioral information. They [social media] also represent a sea of change when it comes to how people and organizations communicate." Online gaming experiences that link to the brick-and-mortar experience are a special opportunity to build data that likewise links the online and physical worlds.

This linking, combined with an industry that is recognized to be at the cutting edge of advanced analytics (as illustrated by the 2012 cross-industry best practice win in advanced analytics by the Seminole Tribe), makes for a special opportunity for gaming to be at the forefront of the online social world. The first step in this process is to source the combined data.

Table 2: Before, During and After Gaming

Before	During	After
The customer can research new games, play games online and review feedback from other players about the gaming experience.	The patron is able to share their gaming expe- rience and possibly event play the online version of games while they are on property.	Players can do further research, receive follow- up marketing offers and play online games dur- ing which they may unlock further stages of the game.

Getting the Data

Speculating about the impact of online gaming is one thing; setting up the tools to measure the impact is another. In today's age of interactions, one of the driving factors is the experiences that the patron had prior to and after their gaming experience. The first step is to source this interaction data—to gain access to those 1 million Clue customers who could be interacting with the game before and after they play in the brick-and-mortar casino. Sourcing the combined data requires a special kind of cooperation between the operator and the manufacturer.

We have stated above that "last century's data breakthrough was transaction data. In other words, the finest grained data was collected by retailers tracking their sales and inventories. ... This century's super data is interaction data." This interaction data can be characterized as data that happens before and after the transaction. Table 2 shows the three stages of interaction: before, during, and after the gaming experience.

Image 2: Total Rewards Portal to Player's Life

Table 3: Critical Aspects of Interaction Data

Predictive	A clear understanding of the interactions a player has before they visit a brick-and-mortar casino can help predict their future behavior. Consider the example of a player whose preference switches from Lord of the Rings™ to Clue. There almost certainly are players who will demonstrate this preference change before they demonstrate the change in the brick-and-mortar casino. Thus we can say this data is predictive.

Multi-property	The interaction data spans multiple properties, and the holder of this data is able to see visitation patterns, length of stay patterns and other transaction data that spans these properties. In the case of Player's Life, this data now spans 1 million players.
Shared Data Ownership	In a world of social media exemplified by Facebook, with its expected 1 billion users as of August 2012,4 consumer data has never been more shared. This means that a player's data is shared between a wide variety of operators; what is not shared are the details of what the player did in the brick-and-mortar property and the entirety of the transaction data. However, with social gaming systems that combine an online and brick-and-mortar experience, these ownership lines are blurring. It should be noted here that Caesar's has its own version of Player's Life at www.playerslife.com/total rewards. In this version, "Total Rewards Members unlock bonus rounds faster with accelerated play." (See Image 2.)
Global	This data is by its nature cross state, cross country and even cross continent.
Location (GPS)	GPS coordinates are a linking factor between brick-and-mortar casinos and social gaming. As was described in part 10 of this series (see the April 2012 issue of CEM), "A customer named Andrew is sitting at a slot machine in his favorite casino, Casino Y. He stops playing for a moment and uses his smartphone to search online for a place to dine." Andrew is generating GPS or locational data that spans both the brick-and-mortar and online worlds, and this gets even more interesting now that Player's Life has a mobile option. Quite simply, Andrew could unlock some more game features while he is dining, or he could even decide on which casino to go to based on the number of available slot machines from the WMS slot finder (see Image 3).

Frequency	In marketing, frequency is one of the core behavioral patterns. To illustrate this, consider the difference between two customers—one who visits once per month versus one who visits once per day, but both spend equal amounts each month. Clearly, it is the frequency of these customers that defines their behavior.
Preference	Game preference is one of the driving factors of understanding gaming behavior on the gaming floor. Social interactions provide clues to the nature of this preference and are likely to give insight into when this preference changes.

Image 3: WMS Slot Locator

A Perfect World

If a perfect world of data cooperation did exist, then the analytical methods available would be squarely in planted in the world of big data analytics. At a high level, the three big analytical questions that should be addressed are outlined in Table 3.

Coexistence or Competition?

Can online gaming and brick-and-mortar casinos coexist? If so, how can this be accomplished? The world is riddled with examples ranging from restaurants and cinemas, where the online world has affected but not diminished the business, to bookstore, which have struggled in brick-and-mortar form. It is impossible to predict the impact of online gaming and what the gaming world will look like in 10 years, but we can say that those who gain and hold the data, learn to understand it and act on it will be able to make decisions that are better informed and more responsive to changes in the market conditions.

Clue is a great game that spans the brick-and-mortar and social media worlds. This exciting product has real potential to help operators and manufacturers bridge the gap between online and in real life, and understand what their patrons do online before and after their gaming experience. It will require special kinds of cooperation, but the results could be remarkable.

1 http://investors.walmartstores.com/phoenix.zhtml?c=112761&p=irol-faq, July 2012
2 http://www.showcase.com/Content/Articles/Retail-Space-for-Lease.aspxm, July 2012.
3 http://www.igt.com/us-en/interactive.aspx, referenced July 2012.
4 http://technorati.com/technology/article/1-billion-facebook-users-now-or/, extracted July 2012.

CHAPTER 15: GREAT GAMES IN GAMING— THE AMAZING BUFFALO

To find the money once again this month, let's dig into the analytical challenges posed by another great game in gaming, Aristocrat's Buffalo. This great game confounds mathematicians and analysts alike, with its amazing ability to continue to generate revenue, despite many versions of this game being nearly 10 years old. Buffalo has been reinvigorated in new cabinets and new gaming platforms but, in essence, it remains the same remarkable game it has always been. Let's explore the analytical implications of such a great game, addressing issues of how to deal with what can only be considered as remarkable data.

Black Swans of Gaming

When trying to understand the world using mathematics, we are often challenged by attempts to predict the outcome of an event or the most successful course of action. A "black swan event"[1] was first defined by Nassim Nicholas Taleb as an event that is so extreme, it redefines the space. According to Taleb, a black swan event is characterized by three factors: "First, it is an outlier, as it lies outside the realm of regular expectations, because nothing in the past can convincingly point to its possibility. Second, it carries an extreme impact. Third, in spite of its outlier status, human nature makes us concoct explanations for its occurrence after the fact, making it explainable and predictable."[2]

We can, then, consider "black swan games" as games that are so extreme in their performance that they are impossible to predict. Buffalo is one such game. In 2012, Goldman Sachs rated Buffalo as the No. 1 game in North America[3], and this is remarkable enough; the fact that Buffalo is a 10-year-old game is astonishing.

Figure 1: Aristocrat's Buffalo Game

If anybody actually understood exactly what the success criteria for black swan games is, there would be a proliferation of hyper-successful games; however, the industry still abounds with less successful games. The problem with pinpointing what will make a game a black swan is in the definition of black swan itself—predicting an outlier is oxymoronic.

In our experience in building predictive models (a truly fun pastime for us authors), we have learned the hard way that models are good at predicting data that follows the central limit theorem but poor at predicting outliers. Yet, in the world of gaming machines, a large portion of the data we care about is outliers. Consider games like Buffalo, as well as IGT's Bombay; these great games are so remarkable and differentiated that their placement and optimization is often critical to the whole gaming floor.

From the perspective of predictive models, it is extremely difficult to model what makes these games successful. Consider the simple example of a game with exactly the same math model that flops on the floor. If math models were the predicting factor of popularity, then we could expect similar customer preference responses on every theme with the same math. But as we all know, that's simply not so on a real-life gaming floor.

Why doesn't predictive modeling work perfectly? It's partly because there is a common misunderstanding that predictive models see into the future. In fact, predictive models take historical trends and tell us what the future will be if these historical trends continue—and they very well might not.

Mankind has been using—and misusing—predictive models far longer than we've been using computers. For example, farmers plan their crops based on historical weather patterns, and when they do this, they are applying a predictive model ("the weather this year should look like ..." + "the soil in this area grows this crop best ..."). The plight of farmers, of course, is when extreme weather that cannot be planned or predicted comes through and destroys the crops. To handle these outliers, the farmer could, and possibly should, purchase insurance to enable survival in the case of disaster. The insurance company, meanwhile, does not predict the disasters but attempts to calculate the likelihood of these black swan events.

Figure 2: Bombay by IGT

The same black swan limitation is true for today's predictive models. While computers give us the ability to take far more variables into account and to understand much smaller nuances in each variable, these models are nevertheless exposed to the same weaknesses as our poor farmers—they cannot account for black swan events. As for the insurance companies, they extensively use computer power to build sophisticated risk models; but in general terms, these models are not all predictive, although they do model the risk in minute detail.

In summary, in today's world of black swan games, predictive models are dangerous but definitely have their place, though they certainly cannot predict outliers. Quite simply, if the model fail, because a successful slot machine is a black swan event.

How Do You Model Artwork?

Let's consider an example of another black swan game, Bombay. In certain regions of the U.S., Bombay is like Buffalo in that it has consistently performed as one of the best core games on a slot floor for many years. Bombay, however, has a "clone" called Sands of Gold that performs vastly less remarkably in multiple denoms and in multiple casinos. Sands of Gold has the same math as Bombay and, aside from the artwork, they are the exact same game. But no matter the situation, Bombay is preferred over Sands of Gold, usually performing 200 percent higher or more.

How does this happen? How do two games that are essentially the same end up deviating so much by having different artwork? And is it the artwork, or did something else drive this—perhaps one game being released before the other?

There are many more examples of this phenomenon of games having the exact same math, but different artwork driving vastly different results. So, in addition to our challenge of trying to predict a black swan, we are now left with trying to include artwork into our models. The approach that we have applied very successfully is a combination of intuition, experimentation and customer behavior. (Examples of intuitive decision making abound in the real world:

Consider how actor Will Smith investigated the marketplace and applied an elementary predictive model to determine which movie he should star in: "First, he gathered the right data—information that was current, accurate, relevant and sufficient to make his decision. Second, he is designed to predict successful slot machines, the model will analyzed it for patterns or insights, and discovered that the top 10 movies included special effects; nine of 10 included special effects with creatures; and eight of 10 included special effects, creatures and a love story. His first two movies, Independence Day and Men in Black, followed that model, and grossed $1.3 billion combined."[5])

The Magical Black Box

But before we get any deeper into our investigation, let's back up for a moment to ask if there even value in trying to predict black swan slot machines, if by nature, they are not predictable. Certainly from the slot manufacturer's perspective, there is immense value in predicting which machines are going to be successful. However, from a slot operator's perspective, is this value reduced?

Let's assume we have a magical "black box" that can predict which of a slot machine manufacturer's games is going to be successful. First, we have to properly define "success." If a game does four times the floor average in its first month, only to decline to half the floor average by the sixth month, this is probably not a successful game. Instead let's define a successful game as one that meets or exceeds the floor average for an extended period of time. Now let's imagine that we have two operators. One operator, Bob, has the magical black box and knows which games are going to be successful. The other operator, Larry, doesn't have this black box.

Bob and Larry both have a bank of six games they need to fill with new product from our manufacturer. Bob has his magical black box, and thus knows that Game A is going to succeed. The problem is, Bob can't order six copies of Game A. If he does this, he will only be appealing to customers who like Game A! While it is going to be a popular game for a long time, Game A is not going to appeal to every single customer. So Bob orders four of Game A and two of Game B to complete the bank. Larry lacks the magical black box, so he orders three of Game A and three of Game B.

Now, after a couple of months, Bob and Larry review the play of their new bank of six games. Bob, with his magical black box, had predicted the right mix of games, and does not make any changes to his floor. Larry, however, sees that he needs more Game A and less Game B, so he is forced to pay for a conversion kit.

For the sake of this example, let's assume that Game A earns $200 per day, compared to Games X, Y, and Z that each make $100 per day. Let's also assume that the cost of conversion kits are $3,000 each. Finally, let's assume it took 60 days for Larry to make his change. With this information, we can put a 1-year value on Bob's magical black box. (See Figure 3)

Figure 3: Summary of Game Experimentation

Bob's 1-Year Net Profit	(4 * 200 + 2 * 100) x 365	$365,000
Larry's 1-Year Net Profit	(3 * 200 + 3 * 100) x 60 + (4 *200 + 2 x 100) x 305 – 3000	$356,000
Incremental Profit		$9,000 (2.5% of Larry's 1-Year Net Profit)

So our magical black box was worth a lift of $9,000 over the course of the first year, and after that first year, there would be no lift at all. So in actuality, it is experimentation on the gaming floor that can best be used to determine the quality of the game.

Review of Customer Preference

We are more interested in optimization metrics than pure game performance. For example, the case study published in the June 2012 issue of CEM6 about Jackpot Wharf showed that even though a "game was performing well ... we can see that customer preference shines a new light onto understanding of the gaming floor. In this example, we should see how players 'moderately interested' in Paradise Fishing have much different product preferences when compared to customers who are moderately interested in Bank A [adjacent bank]. From this, we immediately see that Bank A players are a lot less inclined to play video slot participation products. They are most likely to play house WMS video slot games, some of the older IGT penny titles and Bally's Blazing 7s."

When a Buffalo is a Black Swan

Buffalo is a game that is quite simply amazing. On many gaming floors, it is a category in itself. This incredible game is a perfect example of an outlier that is so extreme that it makes math models based on attributes of games fall apart at the seams. We have discussed that successful games are more like random acts of nature than predictable machines; we can be prepared for them, expect them to occur, and we can even predict their magnitude, but we cannot apply models to determine the performance of a game or if it will be the next great hit. This brings us back to optimization based on experimentation and customer preference. There is little doubt that these methods offer a powerful way of improving bottom-line results.

---

1. Refer to http://en.wikipedia.org/wiki/Black_swan_theory.
2. Extracted from www.nytimes.com/2007/04/22/books/chapters/0422-1st-tale.html?_r=1 July 2012.
3. Refer to http://finance.yahoo.com/news/aristocrat-tops-2012-goldman-sachs-213900500.html July 2012. 4. CEM March 2012, Cardno, Thomas: Where is the Money, Part 9.
5. Source: www.greenbook.org/marketing-research.cfm/will-smith-business-man-06176.
6. CEM June 2012, Cardno, Thomas, Evans, Conklin: "Where is the Money, Part 12: Magnet Games and Paradise Fishing."

CHAPTER 16: OPTIMIZING PARTICIPATION GAMES

One of the biggest questions on any slot floor is how to decide on the right level of participation games. This question of participation is a matter of huge debate and underpins an enormous rift between manufacturers and operators. One party might say that these games bring incremental revenue, while the other might question if participation games are just reallocating revenue that the casino would have collected. Let's dig into this question and outline some real ways that, through customer preference and experimentation, we can discover the true value of a participation game. First, let's introduce ourselves to a couple of operator types that exist when it comes to participation games.

The 2 Percenter Operator

The "2 percenter" operator believes that participation games should be less than 2 percent of the total gaming floor, and in many markets there are very successful operators where this low number is effective. Of course, games such as Wheel of Fortune are still a must, but the performance benchmark needed for the 2 percenters to introduce a participation game is extremely high. If the games are 20 percent revenue share and perform at double house average, then the cost of these games in participation fees is 0.8 percent of the overall gaming revenue.

The 10 Percenter Operator

The "10 percenter" operator has huge numbers of participation games that dominate the gaming floor and are a central part of the overall gaming strategy. An operator with 10 percent of its gaming floor showing participation games pays 4 percent of its overall revenue in participation fees.

The True Cost

As we can see from the wide disparity between 2 percenters and 10 percenters, there is clearly much disagreement in the industry regarding the value of participation games, including wide-area progressives. The disagreement arises from the cost of operating one of these games. Most games have a fixed purchase cost (which can be paid either all at once or in daily increments), but wide-area progressive machines are priced based on a percentage of coin-in. Some of this percentage goes to increasing the progressive meter of the game (which is often more than a million dollars), and some goes back to the manufacturer, but none of it goes to the casino.

To understand how much more expensive a participation game can be, let's look at an example. First, we'll explore the cost as a percentage of revenue. The cost is calculated as a percentage of coin-in, but the true cost to the operator needs to be calculated as a percent of revenue.

To get started, let's look at the overall cost of ownership for participation games compared to non-participation games. Let's assume that the cost to purchase a non-participation game is $20,000 and that over the course of 3 years, the game theme needs to be converted once at a cost of $3,000. Let's also assume the game is winning $150 per day. So the cost to own the non-participation game is $23,000, and over a 3-year period the game will give the casino $150 x 365 x 3 = $164,250 in gaming revenue, for an investment cost (as a percent of net revenues leaving aside cannibalization) of $23,000 / ($164,250 - $23,000) = 16%.

Now, for a participation game, we are still going to assume that it wins $150 per day, but we need to include some more assumptions. First, let's assume the participation cost is 4 percent of coin-in and that the casino hold percentage of the game is 12 percent. Since the game does $150 per day in revenue, with a 12 percent hold it must do $150 / 12% = $1,250 per day in coin-in. Of this, the cost to own is 4 percent or $50 per day. So, over our 3-year period, the cost to own the participation game is $54,750 with an investment cost (as a percent of net revenues leaving aside cannibalization) of 50 percent.

Our investment cost as a percent of net revenues has increased threefold with the participation game, and our gross expense has increased $31,750 over 3 years. Clearly, participation games are very expensive, so operators need to know they are getting value for this expense.

Is the Cost Worth the Return?

Now we get to the controversy. Participation games tend to have average or better than average win per unit—and they are very popular with customers. But are they worth the increased cost to own? To measure this, we need to measure the cannibalization of each participation game. Whenever a slot machine provides a revenue lift (for example, replacing a $100 win per day game with a $300 win per day game), some of that lift is incremental, and some of it is play that shifted from other games. The play that shifted from other games is called cannibalization.

So, to understand if participation games are worth the increased cost to own, we will explore two simple examples: one where the game has low cannibalization, and one where the game has high cannibalization.

For both of our examples, we will compare the value of a participation game that does $150 win per day versus a non-participation game that does $100 win per day. Thus, there is a lift of $50 per day driven by the participation game. We'll also assume that the participation game has a 12 percent hold and costs 4 percent of coin-in. As we calculated previously, the cost to own participation game is $50 per day. In comparison, the three-year cost of $23,000 to own our example non-participation game is just $21 per day. Thus, in the examples below we want to see if we can justify the incremental cost of $29 per day to own the participation game.

The Low Cannibalization Game

First, let's consider a participation game that is very popular and whose customers have shown a propensity not to gamble very much on other non-participation games. In this case, the cannibalization of this game from other games is low. If customers of this game don't play many other games, there cannot be much play shifted from other games to this game. So let's assume that the cannibalization factor is 20 percent—that is, of the $50 lifted from the participation game versus the non-participation game, only 20 percent (or $10) is taken from other slot machines on our floor. In this case, our truly incremental revenue from the participation game is $50 - $10 = $40, which is enough to cover the extra $29 in cost-to-own expenses.

The High Cannibalization Game

In this example, let's consider a participation game that is also very popular; however, the customers who like this game also like to play other non-participation games. In this case, the cannibalization of this game from other games is high. Let's assume that the cannibalization factor is 70 percent—that is, of the $50 lifted from the participation game versus the non-participation game, 70 percent (or $35) is taken from our other slot machines. In this case, our truly incremental revenue from the participation game is $50 - $35 = $15, which not nearly enough to cover the extra $29 in cost to-own expenses. In other words, in this example, we actually lose money on our participation game, despite the fact that it is winning an extra $50 per day over our hypothetical non-participation game.

Market Considerations

Notice that there are many factors involved in this calculation. Cannibalization is one of the key factors, but the floor win per unit is another major factor. The cost to own a non-participation game is fixed, whereas the cost to own a participation game is variable. Thus, for low win per unit per day floors, the increased cost of a participation game is less relative to the non-participation game, making it more likely that it is worth paying the participation fee. For higher win per unit per day floors, the cost to own a non-participation game is very small compared to a participation game, making it less likely that paying the increased expense will be profitable. Thus, market performance affects return on investment.

For example, in a low-revenue market, say $100 per machine per day, the simple economics of revenue sharing games can make a lot of sense. Quite simply, the revenue share may be less than a capital purchase. Daily fee games are a different matter, however, and in high-value markets, say $500 per machine per day, these games are more affordable as a percentage of total revenue. (See Table 1)

Table 1: Revenue Share Cost Comparison

	Revenue Share	Daily Fees
$100 Per Slot Per Day	Low Cost	High Cost
$500 Per Slot Per Day	High Cost	Low Cost

Unlocking Cannibalization

Now let's look closer at participation games' cannibalization, as it is the major determining factor in our calculations of worth today. Unfortunately, trying to determine the effects of a new product on existing products can be daunting. Can we ever measure whether a specific customer's $20 wager was meant for another machine or was in addition to his normal gaming spend on a "typical" trip?

First, let's take a look at the player's behavior using an unnamed but real customer database. When we attempt to determine the growth of a player's worth, one metric we use is their average daily theo (ADT). This metric works fine when you are measuring trip worth over a period of time; for example, to see the effects of your marketing efforts. But if we use ADT as a baseline to measure a player's incremental spend on a single day, the results could be disastrous. In a 3-month sampling of data for the core customer base, we found that only 14 percent of all trips made by players were within a +/- 10 percent of their ADT. More than half of all trips made fell below the 90 percent mark of their ADT and 34 percent of all trips were above 110 percent of the player's ADT. Nearly one-fifth of all trips didn't even meet 25 percent of the player's ADT.

As you can see in Chart 1, the variance in our player's trip theo to his actual ADT is enormous, and we could easily miscalculate whether or not a $20 single session increased or decreased his trip's theo in comparison to his ADT.

Chart 1: Percentage of Trips by Variance to ADT

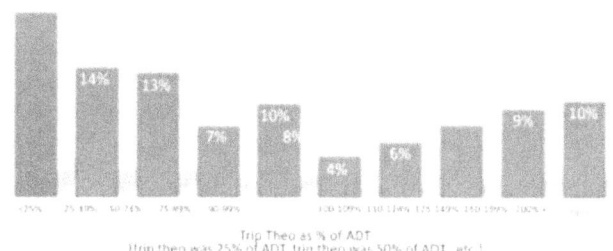

Ghostbusters and Market Basket Analysis

In our experience, most participation games can be shown to primarily cannibalize other participation games. For example, as described in the June 2012 issue of CEM, by performing a preference filter and cluster analysis, we showed that customers who played Ghostbusters in our unnamed database also had a high affinity for penny video slot participation products, both progressive and non-progressive, and penny reel slot progressive participation products. Customers' activity levels in these game categories were high to very high, and no other product category had a high relative impact or activity level. (See Chart 2)

Chart 2: Ghostbusters - Cluster Analysis Results

	Relative Impact	Activity Level
$0.01 Video Slot Progressive Participation	100	Very High
$0.01 Reel Slot Progressive Participation	91	Very High
$0.01 Video Slot Non Progressive Participation	81	High

The preference filter was an extremely important part of this. We limited the cluster analysis through the use of a preference filter to reduce the noise from the numerous trial sessions the product received immediately after its installation. Without the preference filter in place, this noise could incorrectly identify affinity products. For example, thousands of customers tried Ghostbusters in the first 30 days it was on the floor, most never to return to the game. This behavior is completely normal to almost any shiny newly installed game, so among the trial customers were the local video poker and video slot players. Without the preference filter, Ghostbusters players appeared to have an affinity for nearly every product due to the customers' normal trial behavior.

Because Ghostbusters players showed an actual affinity for penny video slot participation products and penny reel slot progressive participation products games, we would expect that these categories would show the biggest effect of cannibalization, as the new games will take a share of the revenue within the affinity product category. Using a simple duplication of purchase model, we could assume that the share would be 2 percent since the new games would equal 2 percent of the unit count in the product category. Experience tells us that new games, particularly participation games, perform at least 30 percent higher for the first month or so following installation before they reach their stabilized performance numbers. It would not be surprising then to see that Ghostbusters has revenue numbers equaling more than 2 percent of our existing product category.

Now the question is: is the new Ghostbusters revenue incremental or did it simply cannibalize the product category?

Imagine that we have a baseline revenue for the affinity product category of $2.5 million. The new product revenue is $115,000. We obviously can't assume it is all incremental revenue, but how much is?

After comparing revenues before and after installation, we see that the existing product revenue shrank by about $100,000. From that, we can estimate the new Ghostbusters product cannibalized 4 percent of the existing product category, and we gained an incremental $15,000, for a 1 percent growth. (See Chart 3) If the new product is significantly more expensive or less expensive than the product it is replacing, we would also need to look at the revenue numbers less fees to determine our net profit change. In this case, the new product was an addition to the floor, not a replacement, and the fees were in line with the product category average.

Chart 3: Revenue Change in Affinity Category After Adding Ghostbusters

This may not always be the case, however. If your cluster analysis shows a high affinity to house games, when using the method explained above, you will have to estimate the cannibalization rate and determine if the new product revenue less fees is greater than the lost revenue to your house games category.

Wheel of Fortune to Move Players

In the Ghostbusters example, we attempted to estimate the incremental revenue and cannibalization of new product to the floor. In the following Wheel of Fortune example, we will show you how moving a bank of participation slots into a high-traffic but poorly performing location increased incremental revenue through impulse buys.

A bank of Wheel of Fortune games was originally placed in a difficult, poor performing area of the casino in an effort to lift performance in the area. It was against a wall in a small lobby area between the main floor and a restaurant but still near the main floor. Although the games performed well above average and the area average increased, we thought we could do better.

Chart 4: Location of Wheel of Fortune on the Casino Floor

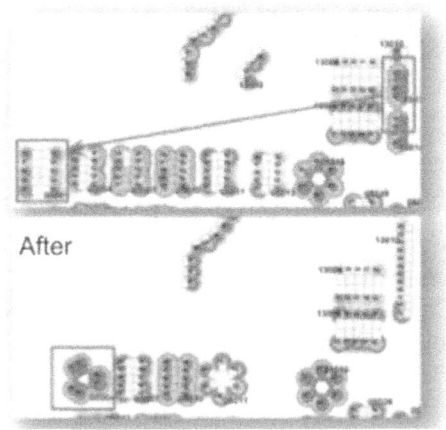

We wanted to place the games in a highly visible high-traffic area, where guests usually queued for the buffet and where there was plenty of foot traffic during entertainment events. Queuing guests and entertainment attendees were not playing the games they were waiting near and walking by, so despite being high traffic, the location was a poor performer. Chart 4 illustrates this move.

Formerly, Wheel of Fortune players were seeking out the game to play it. We were now putting it in full view of guests who were not seeking it out and were not intending to play it. When they did play, it was on impulse and during a time that they were not previously gaming (in line or going to or from the entertainment venue) and therefore did generate incremental revenue.

The entire move affected four banks, including one that was removed from the floor entirely. Even after removing a bank, we saw a revenue increase of 55 percent.

The increase in revenue and player counts happened immediately after the change, as shown in Charts 6 and 7.

Chart 5: Total Revenue Comparison Before and After Change

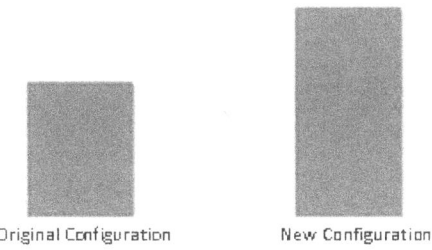

Chart 6: Revenue Trends

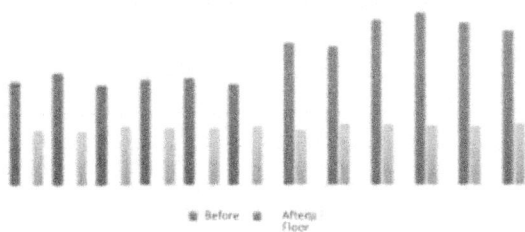

Chart 7: Player Counts

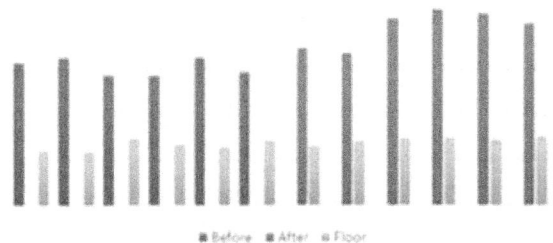

But most interestingly, there was only a 25 percent crossover in the group that played these games in their original location and the group that played these games in their new location. The actual number of handle pulls was flat. And the average bet increased by 42 percent.

Chart 8: Handle Pulls

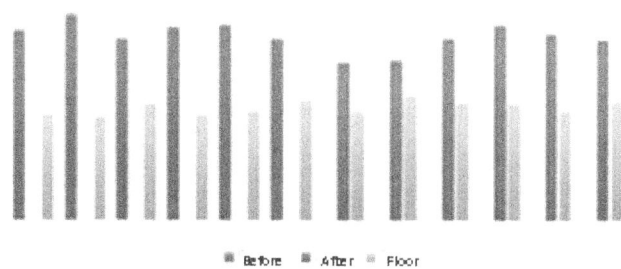

Here we have a case where we have more customers playing the game, but fewer sessions per customers because they are in line or passing by, and a much improved average bet—a much higher percent of max betters than before.

Are Participation Games Making Money?

Analysis of participation games is extremely difficult as we are tackling one of the most difficult analytical questions in gaming. The variation in strategy from the 2 percenters to the 10 percenters, combined with the market and the different pricing models, makes for some real mathematical challenges. That said, as we have shown in our two examples, we can definitely change player behavior and, using preference filters, we can truly see the impact of these important games. As the industry continues to become more competitive, it is the authors' view that these methods become central to good decision making on some of the toughest questions in gaming.

Chart 9: Average Bet

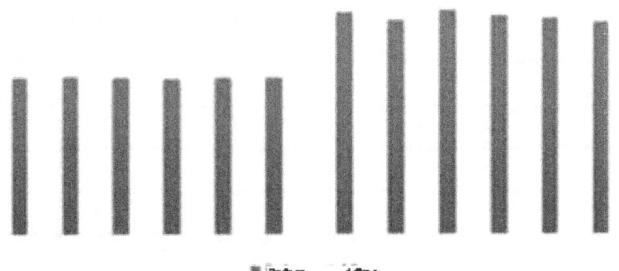

Chapter 17: WILL MACAU BE THE BIGGEST GAMING MARKET IN 2032?

Author's Note: In this article part, we'll analyze the Macau and the Las Vegas markets, but with a twist in time. One of us authors has the privilege of working both in Macau and in Las Vegas. This cross-continent, cross-cultural experience reveals a sharp contrast between the two markets when comparing them in the present day. However, when comparing the Macau of today with Las Vegas in 1992, the similarities are remarkable. This article part digs into these comparisons and then rolls forward 20 years to what Macau might look like if development of gaming in the East follows a similar pattern of growth as it has in the North American gaming market.

Roll back to 1989, before Steve Wynn opened the Mirage and Las Vegas was known to be a wild town; certainly, it was a town without Wall Street investors. The Mirage yielded a new era of gaming in North America: an era backed by Wall Street. In simple terms, gaming seemed like an amazing investment that could support tremendous capital investment, and this capital investment in gaming in Las Vegas culminated with more than $10 billion being invested into CityCenter.

Over the last 10 years, the major corporate gaming companies have largely been located in Las Vegas—even those that have no properties in the Las Vegas market, such as Ameristar and Pinnacle—building not only a center of gaming, but in many ways a center of gaming management. This management center today oversees hundreds of casinos across the world, ranging from clubs in London to a lion's share of the casinos in Macau. It is hard to point to one factor for the mass centralization of gaming management, but the gaming-friendly environment, and enormous and seemingly unlimited revenue potential of the Las Vegas market itself, were certainly contributing factors.

Following this centralization of management, nearly all gaming vendors have either developed a significant presence in Las Vegas or, in fact, have relocated their corporate headquarters to this center of gaming.

Las Vegas has become the center of gaming in North America, and corporate organizations there are now spanning the U.S. and the world with their gaming activities.

As for revenues, in the initial years, it seemed like the Las Vegas market was limitless. You didn't have to look past the amount of capital that was invested to appreciate the management and market potential of the gaming industry. Now, as we look deeper into the Las Vegas data, we will examine two key trends: the average gambling spend and the number of gamblers.

Walking around Las Vegas today, one finds "steel in the ground"—unfinished projects like Echelon and Fountainbleu leaving large swaths of undeveloped land in prime locations on the Strip—as well as shuttered casinos like Sahara. However, the streets have never been more crowded with foot traffic.

The Las Vegas Convention and Visitors Authority collects detailed data on visitors to Las Vegas. As Figure 1 shows, it is quite clear that the gaming market was at one of its highest points in number of visitors per month in 2012. This was validated by the August visitation figures, with 3.34 million visitors in 2012 compared to 3.29 million in August 2011.

Figure 1: Monthly Las Vegas Visitors[1]

However, even given this increased visitation, gaming revenue was down significantly (see Figure 2). Digging deeper into why there was significant growth in visitors and yet a decline in gambling revenue, we can examine the gaming dollars per visit. This critical metric peaked in 2007, but the story does not stop there. Why did it peak? And what can explain the phenomenon of a city showing signs of massive depression, with unfinished and closed casinos, while at the same time looking like the boomtown of Las Vegas in the 1990s? While we won't pretend the answer is simple (or that we possess it), one possible explanation is the impact whales have on a casino's bottom line and the wild volatility that comes with having large amounts of revenues coming from said whales.

But before we look deeper into this, let's draw our comparison to Macau.

In 2004, Sands opened the first "corporate-style" gaming offering in the Macau market. The market has since grown to include six operators, each of which has multiple gaming operations; even the individual "casinos" would be called multi-property operators in the traditional sense. The remarkable thing for these operators is the incredible and seemingly unlimited revenue potential of the Macau market. What is even more incredible is that a relatively small group of casinos were generating $33.5 billion in revenue in 2011. This amount grew in 2012, albeit at a slower rate, and hit $3 billion in September 2012. Those days of heavy growth could be compared to the early days of Las Vegas, and with the enormous population and gambling appetite of China, the market does seem limitless.

Figure 2: Gaming Dollars Per Visit

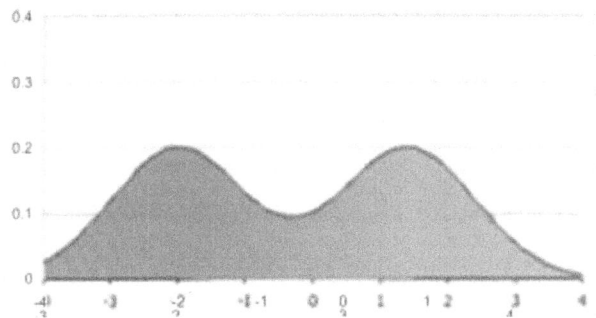

The management teams in this market must be developing highly specialized skills to handle the highly specialized business that exists in this market. It now looks like these management teams are the center of gaming expansion across Asia. In many ways, a presence in Macau is leading the expansion of corporate gaming across Asia. We argue that, like Las Vegas becoming the center of vendors for the U.S. markets, the highly specialized management skills and differences in the markets are likely to lead to a need for vendors to establish a significant presence in Macau as well if they want a strong presence in this growing marketplace.

Gambling Whales

In the casino industry, a "whale" is a player who gambles large amounts of money, but definitions of whales vary wildly. For example, certainly anyone with a million dollar or more credit line would qualify as a whale in any casino, but in Australia, a $50,000 bankroll is also said to classify a player as a whale. Whales typically represent less than 1 percent of a casino's business but can, in some cases, produce 20, 30, or even 40 percent or more of a casino's revenue.

In order to understand mathematically how whales can impact a casino's prospects for future growth, one needs to understand the concept of standard deviation. In probability theory, the standard deviation measures the variation that an event can have from its expected value. We will present two simplified examples of a casino's customer base, one with whales and one without whales, and use the concept of standard deviation to see just how much of an impact whales can have on a casino's bottom line.

Standard Deviation 101

An important rule for standard deviations is called the 68-95-99 rule. This states that for a set of outcomes that have a normal distribution (meaning if you graph the outcomes you get a bell curve), 68 percent of all outcomes lie within one standard deviation of the average (or mean) outcome, 95 percent lie within two standard deviations and 99 percent lie within three standard deviations.

In our example, we will look at two casinos: Casino Grind and Casino Whale. We will assume each casino only has five patrons, and each casino generates $500 per month in gaming revenue. These absurdly low numbers allow us to dig very deeply into the data and truly understand how to measure a casino's expected monthly revenues with the help of standard deviation. More importantly, we will see a stark contrast that is created with the presence of a whale.

Casino Grind's five customers each generate $100 of play over the course of a month. Casino Whale has four customers who each generate $25 of play over the course of a month, plus one whale who generates $400 of play over the same time period. We assume that the likelihood of each customer returning next month is 50 percent, that if they do return they will play exactly the same, and that there is no impact from new gaming customers.

Figure 3: Casino Grind's Number of Possible Outcomes

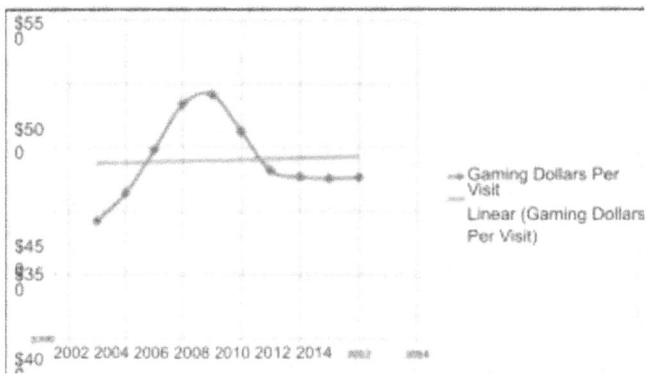

In Figure 3, we see all possible outcomes for Casino Grind. Since each customer has a 50/50 chance of returning, and there are five customers, that leaves a total of 32 = 2^5 outcomes. We then create a histogram of these outcomes, seeing something that fairly well looks like a bell curve. The mean outcome is $250 (since the customers total $500 of play, but each only has a 50 percent chance of returning). The standard deviation of these outcomes is $114. Using the 68-95-99 rule, we see that with 68 percent certainty, the results for the month will be between $136 ($250 -$114 = $136) and $364 ($250 + $114 = $364). In other words, there is a 16 percent chance that the outcome will be below $136 and a 16 percent chance that it will be above $364. In fact, for our example, the probability of being below $136 is 19 percent (6 / 36 = 19%), so the theoretical chance given to us via standard deviations is close to the actual chance found by calculating each possible outcome.

Figure 4: Casino Whale's Number of Possible Outcomes

In Figure 4, we see a very different story for Casino Whale. Looking at the histogram, we see two distinct bell curves caused by the presence (or lack thereof) of the whale. Our standard deviation has increased significantly, from $114 for Casino Grind to $205 for Casino Whale. Instead of there being a 16 percent chance of the revenues being below $136, now we have a 16 percent chance of the revenues being below $45.

So we see how much impact the presence of casino whales can have. Now, we know that casinos don't have five players. However, this type of analysis is repeatable using Monte Carlo simulations. Monte Carlo simulations perform analysis very similar to ours with similar results. But instead of looking at 32 possible outcomes, Monte Carlo simulations look at millions and millions of randomly determined possible outcomes. (The reason they don't look at all possible outcomes is because there are simply too many—if a casino has 100,000 customers, the number of possible outcomes is 2^100,000 which is a number so large it exceeds a 1 with 30,000 zeros.)

Our illustrative example on the effects of whales and high rollers is underpinned by the concept of a bimodal distribution (see Figure 5). If a bimodal distribution exists, such as we think is the case with gamblers and their gambling dollars per visit, then it is, quite simply, mathematically dangerous to look at the average. Our example of Casino Grind and Casino Whale illustrate this danger, and when taken to a larger level, it shows how market analysis for Las Vegas and Macau needs to look both at the total number of visitors and the total high roller market.

Figure 5: Bimodal Distribution[2]

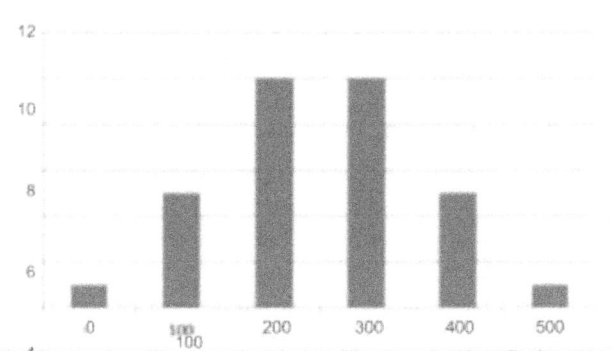

Las Vegas, Macau, and Total Marketplace

It sure is an exciting time in the Macau market, with what seems like unlimited growth opportunities and seemingly endless opportunities to deploy capital. As this marketplace grows, there are three strong follow-on effects for which one should watch. First, watch for management teams in this market becoming the leaders among companies expanding across Asia. Second, watch for vendors establishing strong leadership positions locally to support this center of management. And finally, watch the numbers, as a high volume of customer growth does not necessarily mean growth in revenue.

1. This data is calculated from the sum of Clark County, the Las Vegas Strip, Bounder Highway and downtown. The monthly figures for 2012 are not normalized for seasonality so should be taken as indicative only.
2. http://en.wikipedia.org/wiki/Bimodal_distribution, extracted on October 2012.
WHERE'S THE MONEY?

Chapter 18: SERIES FINALE

Authors' Note: This 18-articlepart journey has shown how we evolved our thinking about player experience-based analytics and how this behavioral analytics model has, in our experience, driven significant revenue improvements. It took us through the great games of gaming and all the way to the growth markets in Asia. This finale will now attempt to summarize the highlights from this series.

There have been three core themes throughout this "Where's the Money?" series: horizontal innovation, the customer experience and analytics. As the industry continues to innovate in many and varied ways, from the systems that tie the games together to the actual gaming devices, this series has provided real-world examples of how analytics is at the core of driving the right value from the customer experience.

Horizontal Innovation

We have defined horizontal innovation as "innovative technologies on the gaming floor that apply broadly across multiple gaming devices." These horizontal innovations are often applied to the whole gaming floor at one time. In other words, the end customer experience is, in many cases, a mixture of influences from many suppliers.

Gaming Standards

We define the gaming product as "the technology and environment that defines the player gaming experience." Key to enabling innovation with this product is the Gaming Standards Association (GSA) and its goal of enabling operators to choose between different suppliers for each component of their gaming systems. This GSA framework enables operators to select from competing products within different horizontal segments of the gaming ecosystem.

The kind of future GSA promises is not conceptually new. In fact, interoperability has been common in the industry for years. For example, customer management systems from a variety of systems vendors all work seamlessly across gaming machines from an even larger variety of manufacturers.

The key is that standards create the platform for horizontal innovation. This style of innovation can manifest itself in unusual ways. For example, the gaming industry has clearly been driven by architectural innovation—just look at the impact that themed properties have had on Las Vegas, where, according to Barry Thalden, "people still flock to the volcano at The Mirage, The Venetian's canals, the Bellagio's fountains and gardens, the New York New York's skyline, and the miniature Eiffel Tower at Paris Las Vegas. And they still come because gambling is fun."1

Gaming Product

Consider the example of Penny Alley, where Silverton Casino used a combination of floor reorganization and horizontal secondary device features to drive significant incremental revenue. This example establishes that customer behavior can be changed by the gaming offering. Contrast the effect of Penny Alley to some new gaming products that Silverton trialed in recent months. These trial games were among the highest-performing games on the floor; however, the calculations of the efficiency (or cannibalization) showed that these games did not add incremental revenue to the property. Of course, these games were returned.

Data and Databases

Data is the core of analytics. The key development in data in recent years is the buildup of third-party databases of information that is, or is likely to be, critical to the operator. For example, Facebook, with its billion users, is accumulating massive amounts of behavioral interaction data on customers, and this information is outside the domain of traditional transaction database information.

Horizontal innovation applies to databases as well as gaming products. For years, slot floor operators have relied on slot performance data to determine what products to place on their floors. Now imagine the power of combining these two data sets, which many casinos have already done or are in the process of doing. From the marketing side, knowing exactly what games are being played by each customer allows for improved segmentation. Coupled with the knowledge of new or changed slot product, marketers can reach out to their customers in far more relevant ways. The same can be said from the slot performance side. Knowing who is playing a particular game can guide product decisions. Have a handful of customers who love Game A? You may decide to keep it, even if it has below average win. Discovered a previously undetected association between Game B and Game C? Make sure to place them near each other!

The Customer Experience

Much of our discussion has focused on the customer. As operators, it is imperative that the customer experience is our highest priority.

Table 1

Players who have the same theoretical win per trip and the same theoretical win per month, but have:		
	Low Theo Win Per Hour	High Theo Win Per Hour
Time-constrained patrons	These patrons spend at a low rate and have limits on the amount of time they can spend at the property.	These aggressive patrons play hard and only stay for short periods.
Time-unconstrained patrons	These players play for extended periods and at a low rate.	These patrons stay for extended periods and spend at a high rate.

Player-Centric Optimization

This is a central theme to our analytics story today. Quite simply, optimization should be done on metrics that the customer experiences, and should be measured by the outcome the business needs. This is defined as player-centric optimization. Table 12 lays the foundations for the different kinds of player optimizations.

Player-centric optimization is defined as improving the player experience in ways that drive the most incremental player net revenue. Expansion of this definition of player-centric optimization results in two kinds of metrics: optimization metrics and outcome metrics.

Optimization metrics measure effects the players can observe. Furthermore, it is generally desirable to optimize these metrics to drive incremental revenue. In short, optimization metrics are metrics that the player notices. Utilization is one metric that players notice—you might hear statements from patrons like "I found my favorite" or "The best games are always occupied when I want them."

Outcome metrics measure effects the players do not observe. For example, the theoretical win per unit per day on the gaming machine. The player experience differs dramatically from the expected or theoretical outcome, and furthermore, the theoretical win per day is an average from a number of different players. Quite simply, players do not experience the spending of other players. Another outcome metric is the slot floor hold percentage, or what is often incorrectly termed the "price" of our games. In our April 2011 articlePart , "Gaming Floors of the Future, Part X: Hold Another Sacred Cow," we debunked the myth that one can increase gaming revenue by simply tinkering with the overall floor hold percentage. Rather, the hold percentage of the whole gaming floor is an outcome metric that is more reflective of the mix of gaming products than it is of the player experience.3

Case Study: Jackpot Wharf

The Jackpot Wharf mini casino strategy4 showed how focused, analytical-driven insight, and operational excellence can result in significant value. The case study followed a number of keys to executing a successful mini casino strategy, including naming the area, spacial management of physical spaces, locational intelligence and spatial data query, and seamless integration into relational data using implicit data relationships,5 all of which were explored in depth earlier in this "Where's the Money?" series.

Great Games of Gaming

From IGT's Wheel of Fortune® to WMS' CLUE™ to Aristocrat's Buffalo, we have studied the games that our customers love. Each of these games provides a different lesson and a different insight into our customers' behavior.

Digging into Wheel of Fortune took us to a study of decay curves and patron growth. One of the key differentiators between decay curve categories (see Table 2) is the ability of a gaming device to attract new players and expand its popularity. Studying the success of Wheel of Fortune led us to a seeming paradox: according to decay curves of most games, we should constantly refresh our gaming floor, but an illustrative example indicated we should keep older games even if they are underperforming. Which approach is correct? In fact, both recommendations are sound. The contradiction exists because we have so many types of patrons, each with their own gaming experience preferences. In the end, the answer lies with the operator's dual mandate: grow revenues and protect revenues.

Table 2: Decay Curves

Decay Curve Type	Description
Never ending	These themes are continuing to attract new players, and manufacturers often continue to release new versions or upgrades. One example is Wheel of Fortune®.
Long-term following	This group of themes has patrons who are super loyal and who will stay with a game for many years. However, these themes are not picking up many new players, so their performance is gradually decreasing. One example in this category is some versions of keno.
Box office hit	These themes are hugely popular when they first appear on the gaming floor, but after a short period, their performance hits a wall. Some movie-based video slot products exhibit this curve.
Never made it	These games never seem to make it. They start low and stay low.

With Clue, we saw the merging of online and offline gaming. What is exciting about Clue is that the player can leave the casino, drive home, and then play online at www.playerslife.com to accumulate features that will change the in-casino gaming experience. While we cannot predict if this model of game play will be a long-term winner, it definitely creates a new kind of analytics challenge and a new kind of data. Clue has real potential to help operators and manufacturers bridge the gap between online and real life, and to help operators understand what their patrons do online before and after their gaming experience. It will require special kinds of cooperation between the two, but the results could be remarkable.

Looking at the great game Buffalo led us to a study of outliers. We described these outliers and some of the mathematical challenges in analyzing them. Our review showed how some games, such as Buffalo, are so extreme in their performance, they are black swans (unpredictable events that bring a major impact). From the perspective of predictive models, it is extremely difficult to model what makes these games successful. Why doesn't predictive modeling work perfectly? It's partly because there is a common misunderstanding that predictive models see into the future. In fact, predictive models take historical trends and tell us what the future will be if these historical trends continue—and they very well might not.

Customer Value and Macau in 2032

It sure is an exciting time in the Macau market, with what seems like unlimited growth opportunities and seemingly endless opportunities to deploy capital. As this marketplace grows, we have illustrated how there are three strong follow-on effects to watch for. First, watch for management teams in this market to become the leaders in companies expanding across Asia. Second, watch for vendors establishing strong leadership positions locally to support this center of management. And, finally, watch the numbers, as a high volume of customer growth does not always mean growth in revenue.

We investigated this final effect in "Where's the Money? Part 17," specifically looking at the impact of whales on Las Vegas revenues. In addition to providing insight into the concept of a bimodal distribution, the study showed how a city can grow in volume of customers but that volume may not lead to growth in revenues. Las Vegas Convention and Visitors Authority6 data clearly showed that at the time of the article's publication, although the gaming market in Las Vegas was at one of the highest points in number of visitors per month in 2012, gaming revenue was down significantly. In fact, this critical metric peaked in 2007.

Analytics

Big Data and Locational Intelligence

Locational data was once strictly the domain of land surveyors7 and geographers. Today, it is generated by nearly every application on any mobile device. Consider this example: A customer named Andrew is sitting at a slot machine in his favorite casino, Casino Y. He stops playing for a moment and uses his smartphone to search online for a place to dine. Andrew is creating GPS data that records his current location and the fact that he is looking for a restaurant. The app provider, to generate advertising revenue, makes this data available to third parties, who in turn use it to place their marketing offers and events. So, Casino Z could be watching for these searches and sending locationally targeted marketing events to Andrew.

Figure 1 shows a screen shot of the Foursquare application Andrew used for his search. As you can see, he is shown several location-based offers in his vicinity—most of which are not at Casino Y.

Figure 1: Location-Based Offers on Foursquare

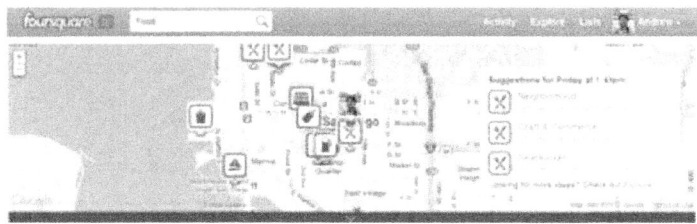

War Room Analytics

Analytics and reporting are completely different things, although on the surface they share many attributes. Table 3 illustrates the difference between analytics and reporting. One of the core ideas of analytics is that it often involves team work with operational components of the business. In the April 2012 issue of CEM, we described the management practice of war room analytics and how it fosters this change.

Big, Complicated, Art

When looking at analytics, we described how analytics is big, complicated and is an art, not a science. In the example in Step 1 (see Table 3), one is pouring over thousands of data elements, trying to find areas of opportunity to improve the customer experience. In Step 3, one is measuring the impact of perhaps hundreds of games and trying to remove the biases created by seasonality, changes in customer preferences over time, and cannibalization. Reporting, on the other hand, is much simpler and smaller, and thus lends itself well to miniaturization to a smartphone or tablet. However, analytics should not be miniaturized.

You heard it here first! Analytics is an art, not a science. Reporting is a science. If you want to know how many widgets you sold yesterday, you take your source system tracking data, ETL it into a data warehouse, push the data into a front-end business intelligence system, then access that system via a computer, tablet or smartphone, and—voila!—you know how many widgets you sold yesterday. However, if you want to know how you can drive incremental widget sales, the task becomes much more difficult.8

Table 3: Analytics vs. Reporting

Analytics	Reporting
Step 1: We want to know what changes we can make to our slot floor to achieve our goal of increasing the player experience in ways that drive incremental gaming revenue. To accomplish this, we first need to know what the optimization metrics (like utilization, spend per hour and devotion) are for every game on our floor. In other words, the desire of the analyst is to increase revenue through optimization.	Step 2: After doing slot optimization, we need to know if we were successful. In particular, were we able to affect win per unit (an outcome metric) in ways that were incremental? At a high level, one can measure the results by looking at, for example, year-over-year gains in win per unit. In other words, the desire of the consumer of the report is to understand the outcome.
Step 3: The report in Step 2 may be skewed by other external factors, like a recession or a booming economy. In order to see past this, one needs to delve deeper into the data and understand the impact made by the changes to the slot floor. In particular, one needs to look not only at the games changed directly, but also at the cannibalization effect on other nearby or similar games. (See "An Analyst's Guide to Slot Floor Optimization," CEM, November 2010.)	Step 4: After digging deeper into the analytics and providing a better measure of incrementality using the analysis described in Step 3, one can summarize this data and demonstrate the effectiveness of the slot optimization. Again, these reports are reporting on the overall results and outcome.

Figure 2: Location of Wheel of Fortune on the Casino Floor

Wheel of Fortune to Move Players

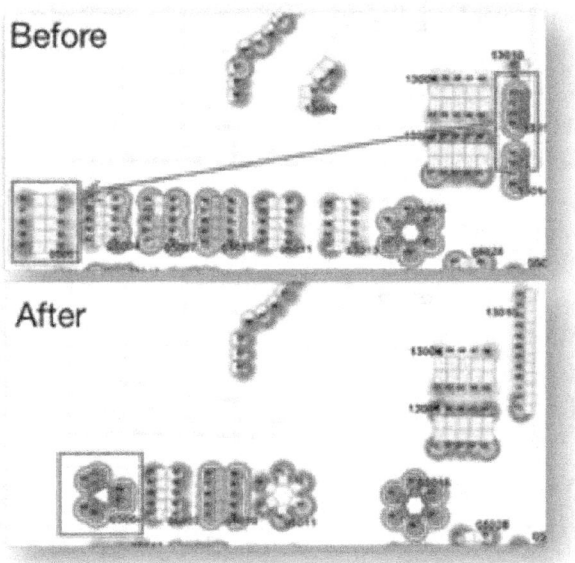

Case Study: Participation Games Optimization

One of the biggest questions on any slot floor is how to decide on the right level of participation games. This question of participation is a matter of huge debate and underpins an enormous rift between manufacturers and operators. One party might say that these games bring incremental revenue, while the other is questioning if participation games are just reallocating revenue that the casino would have collected. We dug into this question and outlined some real ways that, through customer preference and experimentation, we can discover the true value of a participation game.

The "2 percenter" operator believes that participation games should be less than 2 percent of the total gaming floor, while the "10 percenter" operator has huge numbers of participation games that dominate the gaming floor and are a central part of the overall gaming strategy. Figure 2 shows a case study of how participation games can be moved to a high-volume area and that this high-volume area drives incremental revenue. This does not resolve the difference between the 2-percenters and the 10-percenters, but it does show that careful placement of participation games is essential to getting value from these games.

1. "Of Truths and Consequences: How Las Vegas Forgot How to Make Money," Casino Enterprise Management, January 2011, and "Where's the Money? Part 3" Casino Enterprise Management, September 2011.
2. "Where's the Money? Part 4: Gaming Density and Yielding the Floor," Casino Enterprise Management, October 2011, Cardno, Thomas, Gordhan.
3. "Where's the Money? Part 6: Player Experience and Slot Optimization," Casino Enterprise Management.
4. "Where's the Money? Part 8: Player Preferences Learned From Jackpot Wharf, Part 2," Casino Enterprise Management, February 2012, Cardno, Thomas, Evans.
5. "Where's the Money? Part 7: Finding the Money in Jackpot Wharf, Part 1," Casino Enterprise Management, January 2012.
6. www.lvcva.com/stats-and-facts/, extracted October 2012. 7 At least one of the Authors was a land surveyor.
8. "Where's the Money Part 11: War Room Analytics," Casino Enterprise Management, May 2012.
9. "Where's the Money, Part 13: Great Games in Gaming—Wheel of Fortune," Casino Enterprise Management, July 2012, Cardno, Thomas.

Made in United States
North Haven, CT
06 October 2022